Anna Averbakh

Light-weight Experience Collection in Distributed Software Engineering

Logos Verlag Berlin

Bibliographic information published by the Deutsche Nationalbibliothek

The Deutsche Nationalbibliothek lists this publication in the Deutsche
Nationalbibliografie; detailed bibliographic data are available
in the Internet at http://dnb.d-nb.de .

ISBN 978-3-8325-3885-9

Logos Verlag Berlin GmbH
Comeniushof, Gubener Str. 47,
D-10243 Berlin
Germany

Tel.:   +49 (0)30 / 42 85 10 90
Fax: +49 (0)30 / 42 85 10 92
http://www.logos-verlag.com

Light-weight Experience Collection in Distributed Software Engineering

Von der Fakultät für Elektrotechnik und Informatik
der Gottfried Wilhelm Leibniz Universität Hannover
zur Erlangung des akademischen Grades

Doktor-Ingenieurin
Dr.-Ing.

genehmigte Dissertation
von

M. Sc. Anna Averbakh

geboren am 09.08.1983 in Leningrad, Russland

2014

Refernt: Prof. Dr. rer. nat. Kurt Schneider
Korreferent: Asst. Prof. Dr.-Ing. Eric Knauss
Korreferent: Prof. Dr. rer. nat. Robert Jäschke
Tag der Promotion: 01.12.2014

Abstract

In the last decades, many companies have realized that a systematic approach to reuse experience in software engineering is essential for middle and long-term success and competitive advantage. Following best practices based on experience and avoiding past mistakes can help to plan more accurately and to identify new risks. New employees can be trained more easily and experience drain caused by employee fluctuations can be mitigated.

Nowadays, distributed software development has become more common. In a distributed project setting, managing experience is even more crucial than in a co-located project. At the same time, its successful implementation is more challenging. Research has uncovered that efficient and effective work in a distributed project is often complicated by various factors, such as lack of awareness and trust, communicational and organizational overhead, and restrictive information flow policies between project partners. These problems also impede experience exchange and raise the overall effort for software engineers to collaborate. Moreover, sharing experiences is usually not part of the development process and considered additional effort. All in all, experience management initiatives are often doomed to fail due to a lack of participation: the perceived effort to share experience outweighs the received benefit. So far, many researchers in knowledge management have focused on maximizing the (organizational) benefit from the collected experience. However, the effort factor on personal level has been neglected.

This thesis proposes a framework for qualitative and quantitative assessment of *light-weight* experience collection. Light-weight methods primarily aim at lowering the perceived effort and return a reasonable benefit to the experience bearers. This thesis proposes characterizing criteria of light-weight experience collection to assess whether a method is light-weight. This thesis further proposes a measurement system to measure the expected effort and benefit of an experience collection method. This enables assessment of gradations in light-weightedness. To support knowledge managers in choosing the appropriate collection method, this thesis provides a catalogue of strategies from different categories and areas of application in distributed development projects. These experience collection methods are analyzed with the help of the proposed framework. The framework was evaluated in several case studies. Light-weight collection techniques constructed according to the proposed framework are perceived useful and beneficial. The results also indicate a raise of motivation to share more experience.

Keywords: experience collection, light-weight methods, distributed software engineering

Zusammenfassung

In den letzten Jahrzehnten haben viele Unternehmen erkannt, dass ein systematischer Ansatz zur Erfahrungsnutzung im Software Engineering essenziell für einen mittel- und langfristigen Erfolg und Wettbewerbsvorteil ist. Systematisches Lernen aus Fehlern kann zum Beispiel zur präziseren Planung und Identifikation neuer Risiken verhelfen.

Heutzutage ist verteilte Software Entwicklung allgegenwärtig geworden. Forschungsergebnisse haben gezeigt, dass effizientes und effektives Arbeiten in einem verteilten Software Projekt oft durch Faktoren wie eingeschränkte Informationswahrnehmung und Vertrauen, Mehraufwand an Kommunikation und Koordination, sowie restriktive Informationspolitik zwischen den Projektpartnern erschwert wird. Diese Probleme behindern auch den Erfahrungsaustausch für Software Ingenieure, indem sie den wahrgenommenen Gesamtaufwand erhöhen. Dazu kommt, dass Erfahrungsabgabe normalerweise nicht Teil des Entwicklungsprozesses ist und somit einen Zusatzaufwand darstellt. Insgesamt führen viele Initiativen zum Erfahrungsmanagement zu Misserfolg, weil die Mitarbeiter den Aufwand im Vergleich zum Nutzen aus der Initiative als nicht tragbar ansehen. Viele Forscher im Bereich des Wissensmanagements haben sich bisher auf das Maximieren des Nutzens (für das Unternehmen) aus den gesammelten Erfahrungen konzentriert und den persönlichen Aufwand der Erfahrungsträger vernachlässigt.

Diese Dissertation schlägt ein Framework zur qualitativen und quantitativen Messung *leichtgewichtiger* Erfahrungsabgabe im verteilen Software Engineering vor. Leichtgewichtige Verfahren zielen darauf ab, den wahrgenommenen Aufwand für die Erfahrungsabgabe zu senken und gleichzeitig einen annehmbaren Mehrwert für den Erfahrungsträger zu bieten. Um zu entscheiden, ob eine Erhebungsmethode von Erfahrungen leichtgewichtig ist, schlägt diese Arbeit eine Reihe von charakterisierenden Kriterien vor. Um auch Abstufungen der Leichtgewichtigkeit erfassen zu können, wird ein Messsystem zum Bestimmen des erwarteten Aufwands und Nutzens bereitgestellt. Diese Dissertation bietet außerdem einen Katalog von Verfahren aus unterschiedlichen Erhebungskategorien und Einsatzgebieten in verteilter Software Entwicklung an. Diese Methoden wurden mit Hilfe des hier präsentierten Frameworks analysiert. Mehrere Experimente im Rahmen dieser Dissertation zeigen, dass leichtgewichtige Verfahren zur Erfahrungserhebung als nützlich und vorteilhaft von Software Ingenieuren angesehen werden. Die Ergebnisse weisen auch darauf hin, dass sie die Motivation zur Erfahrungsabgabe erhöhen.

Schlagwörter: Erfahrungserhebung, leichtgewichtige Methoden, verteilte Softwareentwicklung

Contents

1. Introduction

1.1. Motivation

Experience and knowledge management is essential in software engineering, as it is known as one of the most knowledge-intensive professions [94]. In the beginning of the 21st century, many large companies recognized the importance of knowledge and experience for software development. Several companies (e.g. Siemens [77], Ericsson [86], Daimler [219], etc.) invested heavily in knowledge management for software engineering problems. Reusing specific software engineering experience by systematic learning is an established way to approach many challenges in software development. Prominent challenges are raising infrastructure and application complexity, knowledge drain due to high staff fluctuation, uncertainty due to volatile requirements and managing correct cost estimation and scheduling [94]. In general, experience and knowledge management helps to strive towards short- and long-term process improvement and competitive advantage [15].

Today, distribution of software-intensive business has become pervasive. It promises cost reduction, proximity to market and customers and access to new workforce [247]. Distribution, however, poses additional challenges to the development process (see Section 2.3) that impede experience sharing in such a project. These are mainly communication and coordination overhead, impaired communication, lack of awareness about information flows, as well as lack of trust. They additionally complicate the development process and increase the workload and stress for software engineers [120, 119]. Due to these reasons, succeeding in a systematic experience exploitation in distributed software engineering is more demanding, but at the same time it is even more crucial than in a co-located setting [82, 83].

As experience resides in the heads of the software engineers, they must be involved and spend their time to share it. To overcome awareness, communication, coordination and trust issues to share a piece of experience can require a lot of initiative and effort from an experience bearer in distributed development. Experience collection activities are usually not part of their everyday project tasks. Software engineers will avoid spending this additional effort. The motivation to share can be extinguished, if the experience bearers additionally do not perceive any benefit from the experience management initiative [33, 110]. As a consequence, many experience and

1

knowledge management initiatives report failure due to lack of participation (e.g. [109, 179, 218, 236, 206]).

Considering this problem, there is a need for a *light-weight* approach to capture experience in distributed development projects. Such an approach should decrease the time to share and effort for the experience bearers but still return a reasonable benefit to them. This thesis provides a framework that enables a knowledge manager to assess the light-weightedness of an experience collection method in a qualitative and quantitative manner.

1.2. Research Questions

This thesis approaches the following main research questions:

RQ1: *What are significant experience sharing challenges in distributed collaborative software development projects?*

RQ2: *How can light-weight experience collection methods be identified?*

RQ3: *How can light-weight experience collection methods be assessed?*

RQ4: *Do light-weight experience collection methods lower the significant experience and knowledge sharing barriers and increase participation in an experience management initiative in a distributed software engineering project?*

In order to answer **RQ1**, I examine the problem domain that drives this thesis: Why is it difficult to collect experience and knowledge in distributed software development projects? I conducted a literature review (see Section 3) on knowledge and experience sharing barriers in distributed projects and identified 63 barriers. The most significant major barriers in a global setting are: No access to knowledge system, lack of awareness about relevance of possessed knowledge and how to submit one's knowledge, lack of trust, content perceived as not useful, no immediate benefit from the experience management initiative and effort greater than benefit. The problems of too much effort and lack of time were emphasized. This literature review provides a confirmation for the need of experience collection methods that focus on effort reduction for the experience contributor.

In order to answer **RQ2**, I provide criteria which a light-weight experience collection method must fulfill (see Chapter 4). It is based on the definition of light-weight and heavy-weight experience collection. The main aim of a light-weight collection method is to lower personal effort for contributing an experience.

In order to enable assessment of light-weightedness gradations of an experience collection method and answer **RQ3** (in Chapter 5), I constructed a measurement system that allows to

2

quantify the criteria of light-weight experience collection. Using these measures, it is possible to measure the effort as well as personal and organizational benefit of an experience collection method. It also provides a scale ranging from very light-weight to very heavy-weight for assessing the light-weightedness (or heavy-weightedness) of an experience collection method.

In order to answer **RQ4**, this thesis reports of several case studies (see Chapter 6). The results indicate that employing a light-weight experience collection method has a motivational effect to sharing more experience and the resulting experiences are perceived as useful. Together with an experience management infrastructure that is specifically tailored to distributed software engineering environments, it can overcome the major experience sharing barriers.

1.3. Assumptions

An important driving force to a successful experience and knowledge management initiative is management commitment (e.g. [196, 214, 179]). Management can enforce experience sharing activities e.g. by rules and policies, reward them or lead by positive example, actively participating in and advertising the initiative. However, rewards and punishments can bring other problems, e.g. loss of quality through gaming the system, which are out of scope of this thesis. Approaches to motivate participation, e.g. rewards, are discussed as related work in Chapter 7. With a strong management commitment, sharing inhibitions – the motivation for this thesis – may not be present or at least not as strong. However, management commitment for experience and management initiatives is often low and it ceases after the initiative has been introduced. Experience sharing is not enforced during the project (see sharing barrier 26 in Appendix A). Light-weight methods can be employed in such a situation. Thus, I assume a low management commitment for an experience management initiative.

1.4. Research Methodology

Software engineering research can be roughly divided into two categories: constructive and empirical research methods [153, p.14]. *Constructive* research method focuses on building and evaluating. Thereby, the researcher identifies a problem and constructs a new solution for this problem. He then measures whether the solution solves the problem. *Empirical* research method focuses on building hypotheses and proving them. The hypotheses are formed by anecdotal evidence and proven or refuted by empirical evaluation – mainly surveys, case studies or experiments [258].

This thesis employs both constructive research as well as empirical research methods. I use constructive research methods to identify knowledge sharing barriers (see Chapter 3) and set up a measurement system for light-weight experience collection. On the other hand, I empirically evaluate hypotheses that prove the helpfulness of light-weight methods and indirectly its measurements, which is an empirical approach (see Chapter 6). As a constructive part, this evaluation demands for new technical solutions. The case studies and experiments in this thesis are exemplary. Its goal is to investigate if light-weight collection methods that lower effort are also beneficial for experience bearers. Thus, the research approaches in this thesis follow the philosophical stance of pragmatism as defined by Easterbrook et al. [93].

Figure 1.1 provides an overview of the employed research methods in this thesis and its structure.

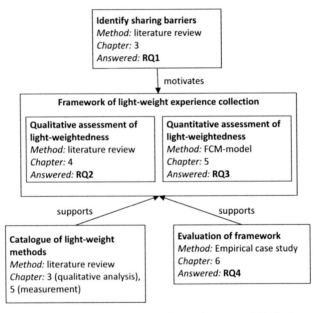

Figure 1.1.: Overview of the methodology and structure of this thesis.

Identify sharing barriers I identify knowledge and experience sharing barriers, from own experience in the distributed collaborative project *e performance*[1] [179]. Taking this rather

[1] *e performance* was a DFG-supported German collaborative project in the area of e-mobility (2009-2012): http://www.audi.de/eperformance/brand/de.html.

anecdotal evidence as basis, I investigate the significance and completeness of these barriers in distributed software engineering (see Chapter 3) based on Kitchenham's method for systematic literature reviews [151]. This answers **RQ1:** *What are significant experience sharing challenges in distributed collaborative software development projects?*

Qualitative assessment of light-weight experience collection In order to operationalize the definition of light-weight and heavy-weight experience collection, I propose criteria derived from this definition and a literature review. This answers **RQ2:** *How can light-weight experience collection methods be identified?*

Catalogue of light-weight methods In order to identify experience collection methods that can be classified as light-weight I conducted a literature review including a search scope and inclusion criteria. The classification was conducted according to the criteria of light-weight experience collection defined in Chapter 4. This provides a basis for a quantitative analysis (see Chapter 5) of light-weight experience collection methods. A knowledge manager can utilize this catalogue as a resource to choose from.

Quantitative assessment of light-weight experience collection Following the FCM-model [172], I derive a measurement system for the criteria of light-weight experience collection. The FCM-model is an approach to derive metrics in order to measure abstract quality factors (F). These factors are broken down into more specific criteria (C), for which metrics (M) can be derived. In this thesis, a constraint for the metrics is that they should be applicable in the planning phase before the collection method has been introduced. This answers **RQ3:** *How can light-weight experience collection methods be assessed?*

Evaluation of framework In order to evaluate whether a light-weight experience management initiative, which was constructed with the help of the framework, is perceived as helpful and motivating by software engineers in distributed projects, two case studies (see Chapter 6) of light-weight experience collection methods were conducted. The methodology of these case studies was explanatory, i.e. "seeking an explanation of a situation or a problem", as defined by Runeson and Höst [205]. I also evaluated whether the system helps to overcome the main experience sharing barriers, if combined with an experience base and experience engineering process. This answers **RQ4:** *Do light-weight experience collection methods lower the significant experience and knowledge sharing barriers and increase participation in an experience management initiative in a distributed software engineering project?*

In summary, the research presented in this thesis is based on a problem identified in practice (my own experience in a distributed project and literature review), a construction of measurements of light-weight experience collection, a literature review of light-weight collection methods and an empirical evaluation of the concepts.

1.5. Thesis Structure

The thesis is structured as follows. Chapter 2 provides backgrounds and relevant concept definitions for experience and knowledge management, cognitive load theory and learning, and distributed software engineering. Chapter 3 presents results of a literature review revealing experience and knowledge sharing barriers in a globally distributed software development. These results motivate a need for light-weight experience collection. The succeeding Chapters 4 and 5 describe the main concepts of this thesis. Chapter 4 qualitatively operationalizes the concept of light-weight experience collection and provides a catalogue of light-weight experience collection methods; the following chapter provides metrics and a scale to assess the grade of light-weight experience collection. The thesis continues with empirical evaluations of the concepts in Chapter 6 by presenting two different light-weight experience collection methods for an EM system. The overall light-weight experience management system particularly supports global software engineering projects and overcomes all main sharing barriers in a distributed setting. Chapter 7 discusses related work and Chapter 8 concludes the thesis and considers areas for future research.

2. Background

In this chapter, I introduce concepts that are relevant to comprehend this thesis. This chapter describes important concepts in the experience and knowledge management field in Section 2.1 and cognitive science in Section 2.2. Finally, I introduce the field of distributed software engineering in Section 2.3.

2.1. Experience and Knowledge Management

This section introduces the domain of experience management (EM), providing a theoretical background for this thesis.

2.1.1. Basic Concepts

This section defines the basic concepts that are fundamental to this thesis: information, knowledge, experience and experience and knowledge management.

According to Davenport and Prusak [76, p.3-4], *information* is defined as

---
Definition 2.1 (Information).
Information is data added with meaning and it is "organized to some purpose".

---

while *data* can be defined as [76, p.2]

Definition 2.2 (Data).
"Data is a set of discrete, objective facts about events."

Information becomes knowledge, if it relates to other information in the head of a person [229]. Davenport and Prusak [76, p.5] define *knowledge* as

---
Definition 2.3 (Knowledge).
"Knowledge is a fluid mix of framed experience, values, contextual information, and expert insight that provides a framework for evaluating and incorporating new experiences and information. It originates and is applied in the minds of knowers. In organizations, it often becomes embedded not only in documents or repositories but also in organizational routines, processes, practices, and norms."

Experience can be seen as a special kind of knowledge. There is no uniform definition of experience. The Oxford Dictionary describes an experience as "practical contact with and observation of facts or events" or "the knowledge or skill acquired by a period of practical experience of something" and "an event or occurrence which leaves an impression on someone" [1]. This thesis adopts Schneider's definition of an experience, which is more specific [216, p. 14]:

Definition 2.4 (Experience).

"An experience is defined as a three-tuple consisting of:

- *an observation;*

- *an emotion (with respect to the observed event);*

- *a conclusion or hypothesis (derived from the observed event and emotion)."*

The main difference of experience in comparison to knowledge is the emotional component. Knowledge is emotionally neutral.

The term *knowledge management* (KM) has various definitions. Dalkir et al. state that there are over 70 good definitions of KM [74, p. 5] from different perspectives, e.g. management, as intellectual asset, cognitive science, information science and more. This thesis adopts a definition that should be consistent with the definition of Software Engineering, which demands "a systematic, disciplined, quantifiable approach" [8]. This thesis uses the definition of KM by Wiig [255], rephrased by Bergmann [36]:

Definition 2.5 (Knowledge management (KM)).

"Knowledge management (KM) is the systematic, explicit, and deliberate building, renewal, and application of knowledge to maximize an enterprise's knowledge-related effectiveness and returns from its knowledge assets."

In comparison to knowledge management, *experience management* focuses on the subset of KM, which deals with the special kind of knowledge – the experience. Thus experience management is defined as (based on Schneider [216, p.7] and Bergmann [36, p.11]):

Definition 2.6 (Experience management (EM)).

Experience management (EM) is a systematic approach to create, collect, store, engineer, maintain and disseminate experience (see Definition 2.4).

The flow of experience in the activities of knowledge management can be viewed as a *lifecycle*. This circle also represents the main EM activities. There are various models and lifecycles in the field of KM. Some major examples are models by Wiig [254], McElroy [175],

8

Rollet [200], Bukowitz and Williams [44] and Meyer and Zack [178], Kolb [154] or Nonaka and Takeuchi's learning model [190]. Dalkir et al. [74, p. 33] present a comparison of the first five models. All these cycles include a creation, acquisition, transformation, storing and distribution of knowledge or experience.

In this thesis, for the goal to convey a general flow of experiences, the model presented by Schneider [216, p. 143] is sufficient. The other mentioned life-cycles are either too sophisticated for this purpose (e.g. Nonaka and Takeuchi's model [190]) or do not consider the emotional component that distinguishes an experience from "neutral" knowledge (e.g. Kolb's learning cycle [154]). The life-cycle this thesis refers to is displayed in Figure 2.1.

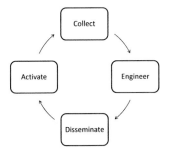

Figure 2.1.: The experience life-cycle based on Schneider [216, p. 143].

Collect During this step, experience is collected from experience bearers (directly or indirectly through documents) and stored in the experience base.

Engineer Afterwards experience artifacts are engineered to convert subjective experience into reusable recommendations (e.g. as checklists).

Disseminate The engineered recommendations should be actively distributed to all interested parties.

Activate Disseminated experiences may may be confirmed, refuted or limited to a more specific context through application. This can lead to new experiences or change old recommendations. They, in turn, can be collected, which starts a new cycle.

This thesis mainly focuses on the *collection* step, but also influences activation and engineering. Experience dissemination is not a subject in this thesis. The concepts relevant for experience collection are introduced in the next section, while experience engineering is described in Section 2.1.3.

2.1.2. Experience Collection Domain

Before illustrating the experience collection methods, a basic understanding of the terminology and their relations is needed. The domain model in Figures 2.2 and 2.3 interrelate the terms that are relevant in the context of experience collection[1]. In the following, each concept is defined in more detail.

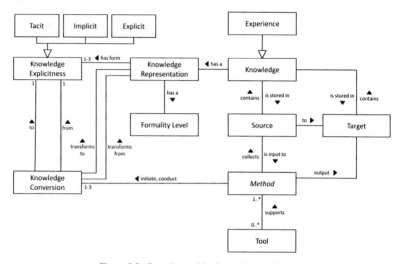

Figure 2.2.: Domain model of experience collection

As already defined in Section 2.1.1, *experience* is a special kind of *knowledge*. A piece of knowledge always has a *representation*. It is the form that knowledge can embrace. Free text, ontology or use cases are typical examples of knowledge representation. Each knowledge representation has a *formality level*. The more formal the representation, the more easily it is to automatically analyze and process the knowledge. Each knowledge representation has one of three cases of *knowledge explicitness*: *tacit*, *implicit* or *explicit*. Tacit knowledge, coined by Polanyi [194], resides unconsciously in the head of a person. Implicit knowledge is consciously kept in the head of a person. Explicit knowledge has been externalized, e.g. as text in a document, audio recording or ontology. Knowledge always resides in a *source* and flows to a *target* if shared with the help of a method. In contrast to knowledge representation, source and target

---

[1]The domain model in Figures 2.2 and 2.3 is part of the contribution of this thesis. For better understanding, it is presented in this chapter.

are physical artifacts containing knowledge. A source can be a specific person with a thought as knowledge representation. A target could be a specific OWL ontology file [5] that contains knowledge represented (structured) as an ontology. An ontology is "a set of representational primitives with which to model a domain of knowledge or discourse" [107].

A *method* to collect experiences can be specialized into elicitation or extraction method (Figure 2.3). A method is called *elicitation method*, if knowledge to be collected comes from a person's mind. If knowledge is already externalized, e.g. into documents, audio or video sources, the collection method is called *extraction method*. Many collection methods employ several methods. These are called *composite methods*. Normally, the "component" methods are not equally employed in a composite method. One method stands in the foreground as the *main method* of collection and the other methods are *secondary* or *supplementary* methods. An illustrative example is presented further down.

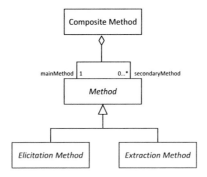

Figure 2.3.: Detailed domain model of a collection method.

A method can be supported by a *tool* (Figure 2.2). A tool for knowledge collection must support at least one method. This implication means that a knowledge management project should only utilize tools that serve the purpose to support a method. A method initiates or conducts one to three *knowledge conversions*. According to Nonaka and Takeuchi [190], these are socialization, externalization and combination[2]. During the process of knowledge conversion, knowledge changes from one knowledge explicitness to another. Along with the form, knowledge representation is usually transformed into another one as well. Example 2.1 brings together and illustrates all terms of this domain model.

---

[2]The fourth conversion *internalization* is left out intentionally. The focus of this thesis lies in experience collection and not dissemination.

Example 2.1 (Experience collection domain).

Scenario: *Automatically extract free text annotations from project documents, displaying them in a shared experience base (e.g. a Wiki) for collaborative commenting and search.*

The method is a composite method: 1) The main method is to extract annotations into a shared experience base and 2) the secondary method is to elicit comments added to the extracted annotations. The first is an *extraction* method because the source is a textual document and the second is an *elicitation* method, since the sources are people.

For the main method, the sources of knowledge are the specific *files containing the annotations*. The target is a *shared experience base*, where the annotations are extracted. Knowledge representation of the source is annotations (or more general: textual form). The representation of the target is free text, too. The formality of the source representation is informal. Although annotations can be automatically extracted, they are written as free text. To extract valuable knowledge from annotations is not trivial. The target can have any formality, depending on the setup of the shared experience base. The knowledge conversion happening in this method is *combination*: from explicit knowledge to explicit.

For the secondary method, the source is *people commenting the annotations* and the target is a *shared experience base*. Knowledge representation of the source is *conscious thoughts*, which is *implicit knowledge*. Knowledge representation of the target is free text. Knowledge conversion is therefore *externalization*. The formality of the source is informal: the target can have any formality.

2.1.3. Experience Engineering

Based on Schneider [216, p. 145], this thesis defines experience engineering as follows:

Definition 2.7 (Experience engineering (EE)).

Experience engineering (EE) is an activity that improves experiences or related material by applying systematic procedures, similar to engineering.

The main artifacts in the experience engineering process are defined below and displayed in Figure 2.4. These are a *raw experience*, a *recommendation* and a *best practice*. The definitions for a raw experience (Definition 2.8) and best practice (Definition 2.10) are based on Schneider [216], while recommendation (Definition 2.9) is based on Averbakh et al. [23].

Definition 2.8 (Raw experience).

Raw experience is experience that has been collected from experience bearers but has not been engineered yet. Usually it is highly subjective, confidential, unreliable and incomplete, concerning the aspects that define an experience (see Definition 2.4).

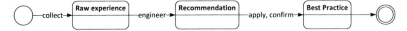

Figure 2.4.: Main artifacts in the experience engineering process.

Definition 2.9 (Recommendation).

A recommendation is a reusable procedure to follow, engineered from raw experience. In contrast to a best practice (Definition 2.10), a recommendation has not yet been applied and confirmed as useful in practice.

Definition 2.10 (Best practice).

A best practice is a recommendation that has been applied at least once and confirmed as useful in practice.

Experience engineering can include the following tasks. They are also based on Schneider's definition [216, p. 146] but adopted to distributed software engineering settings [23].

1. Go through (i.e., read, look at, watch or listen to) collected material.

2. Identify and sort out potential raw experiences from other information ("noise").

3. Clean, anonymize, rephrase, and filter raw experiences.

4. Extract remarkable and recurring passages by aggregating and comparing raw experiences.

5. Categorize (e.g. tag, index, describe shortly) experience extracts.

6. Harmonize experiences in size and granularity. They should preferably be of medium size and mixed granularity (see Section 5.3.3).

7. Derive recommendations or procedures to follow from experience extracts.

8. Select an appropriate presentation style for recommendations.

9. Maintain experiences: promote recommendations to best practices and track outdated experiences. This step can fall into experience maintenance or evaluation. Usually, the decision to promote, demote or edit a recommendation is made by experience bearers and the experience engineer processes this feedback.

2.1.4. Experience and Information Flow

As introduced in Section 2.1.1, experience and knowledge has a life-cycle, where it circulates and evolves. To illustrate experience or information flows between actors and systems, I employ the FLOW notation [231, 229], which is explicitly designed to display information and experience flows. These diagrams summarize processes, general roles, as well as information and experience flows that are part of the specific method. Notably, FLOW diagrams make clear in which case experience is captured directly or indirectly (through e.g. documents) from the bearers. Figure 2.5 gives an overview of the FLOW syntax.

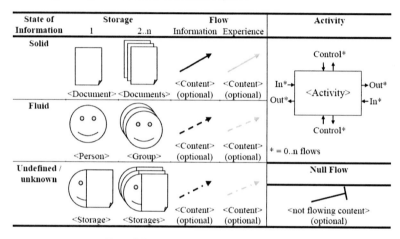

Figure 2.5.: The FLOW syntax definition [231].

2.2. Cognitive Load and Learning

How does the mind and cognition work and how do we learn? According to the *Cognitive Load* theory [240] there are two mechanisms of learning: "schema acquisition and the transfer of learned procedures from controlled to automatic processing" [240]. Sweller defines a *schema* as "a cognitive construct that organizes the elements of information according to the manner with which they will be dealt" [240]. In other words, people do not learn the concrete things they encounter but more abstract schemas with characteristics. A person would not remember the particular tree he actually saw, but rather the main elements of a tree (e.g. leaves, branches and

14

color). The same mechanism applies to dealing with problems. People categorize problems into schemas, i.e. problem classifications. This way a person's mind can efficiently deal with the infinite variety of specific problems and expressions. Example 2.2 illustrates how people learn and think in schemas.

Example 2.2 (Cognition).

An experienced developer, who knows many programming languages, would understand a code in a new language yet unknown to him. Having internalized the schema of programming languages like the commonest constructs (variables, statements, conditions, loops, etc.) and how they should usually look like, he will most probably understand the general task the code fulfills. Without a language schema, the developer would have to learn the language from the very beginning to understand it.

The concept of acquiring schemas knows only two states: a person either has acquired a schema of something or he has not. Usually people learn in gradations. A novice developer or student, who has just learned his first programming language, would not be able to easily transfer abstract language concepts to other languages like an experienced developer. He would spend considerable effort and thought to do this task. This way of processing information is called *controlled* processing. In contrast *automatic* processing occurs if it is subconscious. In the knowledge management field, this resembles to tacit knowledge (see Section 2.1.2), which also resides in the unconscious. According to the *Cognitive Load* theory, engaging in complex activities with controlled processing requires a high cognitive load. The term cognitive load can be defined as [240]:

Definition 2.11 (Cognitive load).

Cognitive load of a task is the load on the internal information processing capacity of the individual needed to solve the task.

The human information-processing system for computer-supported work[3] consists of two separate dual channels: 1) *auditory / verbal* for processing auditory and verbal input and 2) *visual / pictorial* for processing visual input and pictorial representations [171]. Each channel has a limited capacity. Only a limited amount of cognitive processing can take place in both channels at one time [171, 56]. An example would be the well known observation that an average human mind can only process $7+/-2$ information items at a time [181].

In the domains of HCI and Usability Engineering, cognitive load is an important field of attention. Cooper et al. differentiate four types of work that should be minimized in non-entertainment products or processes to increase user effictivity, efficiency and satisfaction [68, p.151]. Experience collection can be considered as such a process:

---

[3]Smell and taste are considered irrelevant in the context of computer-supported work.

- *Cognitive work:* Comprehension of system behaviors and text

- *Memory work:* Recall of system behaviors, commands, locations of data and objects, and other relationships

- *Visual work:* Figuring out where the eye should start on a screen, finding an object among many, differentiate visually coded objects (items with different color)

- *Physical work:* Keystrokes, mouse movements, gestures, input mode switches, number of clicks needed to navigate

In this thesis, I combine these aspects into the subsuming term *cognitive load*. I consider terms like *mental load* or *cognitive workload* [168] synonyms to *cognitive load*.

2.3. Distributed Software Engineering

Distribution of tasks and processes always implies collaboration with other parties. Collaboration in general can be viewed from two points of view: as legal entities and as locations. Of course, characterizing a collaboration, both points of view have to be considered. Figure 2.6 sums up the the various aspects of distributed work in general, which is directly applicable to software engineering. From the legal point of view, collaboration can take place within a com-

Legal entities   **Locations**

Figure 2.6.: Different aspects of distribution (based on [246, Fig. 2]).

pany or with another company. In both situations the partners can be located either all within or outside of the country. In case of a collaboration with partners outside of the country, the location can be far or near [246].

Literature on distributed projects does not present a uniform terminology for this topic. There are various terms like offshoring, onshoring, inshoring, outsourcing, insourcing, etc. [246]. Even global software development is not unanimously used throughout literature. Often *global* is used to denote international collaboration outside of the country, i.e. around the globe (e.g. [121, 120, 183]). Šmite et al. however, attempting to introduce a common taxonomy in this

16

field, define the term more generally as "[d]evelopment of a software artifact across more than one location" [246].

In this thesis, I use *distributed* software engineering as a general notion of developing across multiple sites to avoid the connotation of globalization. I will use the term *global* or *globally distributed* software engineering to denote the special situation of a distribution far outside of the country. For this thesis, I focus on a collaboration with another company and assume that the collaborating partners may produce knowledge and experience that is sensible and must not be disclosed to partners (e.g. through NDAs). The concept of light-weight experience collection does not explicitly consider trust and information disclosure issues, but I discuss it as a major experience sharing inhibitor in Chapter 3. Further, this issue is also addressed in the experience enigneering process in Section 2.1.3. Also the applications presented in Chapter 6, especially the presented experience base with rights managements (in Section 6.3), specifically addresses this issue and presents solutions to overcome it.

Generally, *distributed* software engineering has a number of characteristics [183]:

- *Multisourcing:* Multiple distributed collaboration partners are involved in a joint project.

- *Geographic distribution:* Project partners are located away from each other.

- *Contextual diversity:* There is a diversity in process maturity and work practices of each project partner.

In the special situation of *global* distribution, more factors come into play[4] [183]:

- *Political and legislative diversity:* Collaboration across borders can have aggravating effects due to political threats or threats associated with incompatibility of laws.

- *Socio-cultural diversity:* Different cultures of partners bear different levels of social, ethnical, and cultural fit.

- *Linguistic diversity:* Partners can have different levels of language skills.

- *Temporal diversity:* Global partners can be situated in different time zones with differing working hours.

These factors usually lead to challenges that are specific to distributed projects [121, 119]. These are:

---

[4]These factors, except *temporal diversity*, also apply to near-shore multi-organizational distribution, though.

- *Impaired and less effective communication:* Distributed projects usually suffer from less frequent and less effective communication. This effect is due to temporal, socio-cultural and geographic distance. Because of geographic distribution, there is much less direct (i.e. face-to-face) communication and more indirect communication, e.g. through instant messaging, email, telephone or video-conferencing. These types of communication have a higher latency (time until a response is received [229]) and narrower bandwidth (amount of data that can be sent by the communication medium [229]) in comparison to speaking face-to-face. Communication latency has a negative influence on communication efficiency [229, p.157]. Different time zones impede communication even more by limiting meeting times. In the worst case with no working hours overlay, only asynchronous communication is possible. Cultural and language problems also lead to understanding and interpretation problems during communication.

- *Lack of awareness:* Project participants at different sites often share little context: They have little knowledge what their colleagues at the other location are doing every day, if they are available for contact and what their current concerns are. They may not become aware of changes. Lack of context makes it difficult to initiate contact and often leads to misunderstandings and miscommunication. Of course, impaired communication also aggravates this situation.

- *Incompatibilities:* Distributed partners often have different work practices, processes, habits and corporate culture. These practices can be incompatible leading to problems, confusions and misunderstandings. An example is sharing negative experience and failures with partner companies, which can be a common practice at one site but be a very delicate matter at the other.

- *Lack of trust:* Dispersed projects often suffer from a less trustful atmosphere between partners. Lack of trust can be amplified by partner companies, if they set up NDA agreements and increase behavioral control. Additionally, different cultures, languages, little direct communication and poor socialization leads to a lack of trust among the teams [183].

In the next chapter, I point out how these challenges of distributed software development affect and inhibit knowledge and experience sharing.

3. Challenges of Experience and Knowledge Management in Distributed Software Engineering

To determine which knowledge and experience sharing problems to approach, this chapter provides an overview of sharing barriers. The research method of this analysis is a literature review based on Kitchenham's approach to conduct systematic literature reviews [151].

This review aims to answer the first main research question:

RQ1: *What are significant experience sharing challenges in distributed collaborative software development projects?*

From this research question, more specific questions can be derived, which drive this review:

RQ1.1: *What is the state-of-research on experience and knowledge sharing barriers?*

RQ1.2: *Which of the barriers are specific to globally distributed collaborative software development projects?*

3.1. Review Methodology

To create an overview of the experience and knowledge sharing barriers in global software engineering, an extensive literature review was conducted. First, the keywords for the search query were set (Table 3.1) to (A1 OR A2) AND (B1 OR B2 OR B3) AND C.

Table 3.1.: Keywords for the search.

A1—knowledge sharing	B1—barriers	C—software engineering
A2—experience sharing	B2—impediments	
	B3—issues	

To perform the search, Table 3.2 displays the decisions made.

As inclusion criteria, publications were considered relevant if they report on knowledge or experience barriers. This includes general personal sharing barriers or barriers that concern groups

Table 3.2.: Decisions for the literature review. The values in brackets represent the number of
retrieved results in the database.

Searched databases (# results):	Emerald (55), IEEE Xplore (0), Springer Link (102), Sci-enceDirect (2), Wiley InterScience Journal Finder (2), ACM Digital Library (994), SAGE Journals (28) and Google Scholar (19700 found, 200 used)
Searched items:	Journal articles, workshop papers, and conference papers.
Search applied on:	On title of the publications written in English language and the keyword "software engineering" on the text.
Language:	The selection was limited to publications written in English.
Publication period:	A period was not explicitly defined.

of people (social and cultural issues). In a first review iteration and motivated by own experience in a distributed collaborative project [179] it became clear that many barriers are not specific to distributed software development. In fact, almost all barriers except those concerning communication, distribution and trust issues were mentioned in literature without the specific focus on distributed For this reason, this review is not restricted to distributed settings. It poses new challenges to experience and knowledge sharing but other barriers have influence on distributed projects as well or are amplified by distribution.

Some publications were also found through a recursive search, as some of the initially found sources named the barriers and referenced to further sources. Some of these sources could not be directly found with the used keywords. For the Google Scholar search, I only used the first twenty pages (200 results).

3.1.1. Data Collection and Analysis

Altogether, 78 relevant publications and 63 barriers were identified. From each publication relevant to this review, factors and barriers were collected into a table. Afterwards similar barriers were consolidated and grouped according to their barrier category. If the publication named a causal relation between the factors, this relation was drawn. These were not many. Most causal relationships, however, I only indirectly deduced from the given literature. A complete overview of the identified barriers and sources can be found in Appendix A.

From the collected barriers I considered only those relevant that are related to distributed software engineering projects (including global distribution) or directly influence them. I identified many barriers as refinements or instantiations of more general ones, e.g. *knowledge sharing costs too much for the management* being a generalization of *management spends too little per-*

sonnel resources on the EM initiative. I did *not* consider barriers that were not directly related to or directly influenced knowledge *sharing* of individuals or groups. An example of such an issue is having an *incompetent knowledge management leader* [13]. This problem does not directly influence knowledge sharing. Of course, many or even all of the found barriers can also lead to a failure of the EM initiative as a whole.

3.2. Identified Experience and Knowledge Sharing Barriers

To answer **RQ1.1:** *What is the state-of-research on experience and knowledge sharing barriers?* Appendix A lists all identified barriers. The barriers are ordered by their category containing the sources. The choice of categories is based on categories I often encountered in the reviewed literature, e.g. in [196, 27, 179, 237]. The chosen categories with an explanation are displayed in Table 3.3.

Table 3.3.: Categories of knowledge sharing barriers.

Category	Explanation
DSE	barriers specifically related to (globally) distributed (software engineering) projects
organizational	barriers influenced by organizational shortcomings
social	socially, psychologically or culturally influenced barriers
technological	barriers related to technology (tools)
process-related	barriers caused by wrong processes related to the experience lifecycle

Table 3.4 provides an excerpt of experience and knowledge sharing barriers that have been most often references in the reviewed literature. The overview shows that, beside trust and confidentiality-related barriers, lack of time (ID:36), bad workflow integration (ID:50), excess of effort (ID:51) and the imbalance of perceived effort and benefit (ID:52) are often named in the reviewed literature. *Not enough time* (ID:36), having the highest reference count, indicates a need of experience collection approaches that minimize the time to share and effort for the experience bearer. The sources to these barriers can be found in Table A.1 in Appendix A, while Table A.2 in this appendix lists all barriers ordered by the frequency of their references in the reviewed literature. The barrier IDs in Table A.2 can be used to look up literature references in Table A.1.

To answer **RQ1.2:** *Which of the barriers are specific to globally distributed collaborative software development projects?* the diagram in Figure 3.1 displays a selection of barriers that are specific to (globally) distributed collaborative (software development) projects. I filtered those

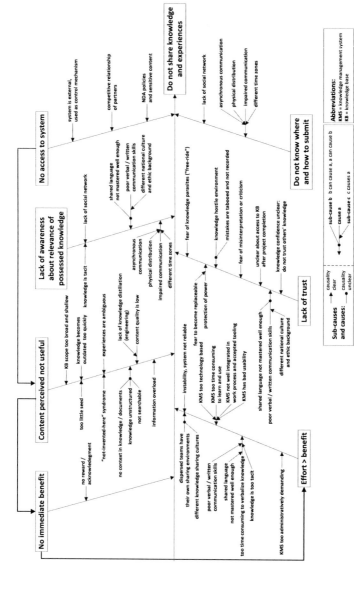

Figure 3.1.: Experience and knowledge sharing barriers in a cause / effect relationship in a (globally) distributed project.

Table 3.4.: The top twelve experience sharing sharing barriers from Table A.2 in Appendix A with the highest number of references.

ID	Category	Barrier	References in Literature
36	social	not enough time	15
37	social	protection of competitive advantages / power, fear to become replaceable	13
1	DSE	NDA policies and sensitive knowledge (confidentiality)	12
50	social, tech.	does not fit into workflow	12
3	DSE	different knowledge sharing cultures	11
9	DSE, social	lack of trust	11
40	social	no reward / acknowledgement	10
42	social	lack of perceived self-efficacy (awareness about relevance of possessed knowledge)	10
51	social, tech.	too much effort	10
26	org.	not enough management imposition / commitment / interest	9
41	social	evaluation apprehension (fear of misinterpretation or criticism of one's knowledge)	9
52	social, tech.	more perceived effort than benefit	9
...			

barriers that were either mentioned in relation to distributed or collaborative (SE) projects or were directly influenced (amplified) by physical distribution. Most of the organizational knowledge sharing barriers were omitted in this overview, as the topic of this thesis is to mitigate sharing barriers on a personal level. Organizational issues like *lack of management imposition* or a *too big size of business units* prohibiting effective sharing, etc., are considered unchangeable in the scope of this thesis. These barriers are also listed in Appendix A. The context of this thesis is *distributed* software development projects. The term *global* software development is a more specific notion, which implies different locations over the world with challenges like different time zones, languages and cultures. Symptoms of distribution can already occur with being about 30 meters apart [119, 17]. This thesis thus does not focus on cultural issues and time differences, but the analyzed solutions and the general approach of light-weight experience collection can also be applied on a globally distributed project. It must only be adapted to the current culture. More specifically, this thesis focuses on distributed multi-firm collaborations, which raises the problems of data access policies and competitive relationships.

As an interpretation of the literature and own judgment, I put the barriers in a causal relationship. I identified six major causes to the knowledge sharing problem. Each major cause can be caused by minor causes, which in turn can have sub-causes. Some of the relations cannot be definitely stated as causal (the sub-cause could also be an effect to the barrier it causes), but only as associations. These relations are denoted by a point instead of an arrowhead in the diagram. There are also causal connections between major causes, which is indicated by the arrows. Some barriers, e.g. those causing *impaired communication*, can be found as causes for several major barriers. These are mostly issues common to distributed projects [121, 120, 122, 182, 250].

In the next chapter, each barrier and its sub-barriers from Figure 3.1 is described in detail.

3.2.1. Barriers in Detail

This section explains each barrier from the overview in Figure 3.1. The barriers are ordered according to their causal structure in the diagram. The headings are the major identified barriers. For each barrier I list its ID (in brackets) referring to the ID in Appendix A. This should enable a quick lookup of the literature sources. Some barriers are marked with a *(B)* meaning that these barriers are not directly experience sharing barriers, but *basic* challenges in distributed software engineering as introduced in Section 2.3. However, they have a direct impact on a knowledge and experience sharing barrier and are therefore presented in this listing. The order in which the barriers are presented corresponds to their placement in Figure 3.1, not in Appendix A.

No Access to System (ID:16)

In this issue category I approach the problem that collaboration partners may have restricted access to the shared knowledge. This issue and the causing sub-issues are special to collaborative projects, where multiple companies are involved. The causes of this barrier are:

System is external, used as control mechanism (ID:7): This is an issue if the knowledge base is hosted at one company, e.g. the major partner, who can regulate the access at will. Other partners are uncertain what happens with the knowledge after the project ends. With this uncertainty in mind, they may withhold their knowledge.

Competitive relationship of partners (ID:2): Collaboration partners can be competitive outside of the collaboration. This competitive relationship hinders the flow of knowledge between them. This often affects access rights to knowledge created during the project, especially if the system is external and hosted by one of the partners.

NDA policies and sensitive content (ID:1): Non-disclosure agreements (NDAs) due to, e.g. a competitive relationship between partners, may hinder knowledge flow to other partners in the collaboration.

Lack of Awareness About Relevance of Possessed Knowledge (ID:42)

This issue happens if the project participant does not know if his knowledge is relevant to other project participants. This is a common problem, not special to collaborative projects, but can be amplified by the physical distribution. The causes of this barrier are:

Lack of social network (ID:33): A social network in this context is "[a] network of social inter- actions and personal relationships." [7]. Not having a friendly relationship to colleagues at the company means that a person may become isolated and not know who to ask for help. This can be an issue in local projects, if communication between colleagues is not frequent. It is amplified in (globally) distributed projects due to less communication. Knowledge is not exchanged, if project participants do not know anybody who may need the knowledge.

Poor verbal / written communication skill (ID:12): This is an issue that is not special to dis- tributed software engineering, but can be amplified through cultural diversity. Some peo- ple cannot articulate their knowledge well in a verbal or written manner. The sub-causes are:

  Different national culture and ethnic background (ID:13): Distributed project teams of- ten consist of people from different countries. Different cultures often means dif- ferent values, views and beliefs, which hinder communication and knowledge ex- change.

  Shared language not mastered well enough (ID:14): In multi-cultural distributed projects, participants may not speak a common language (most often English) well enough to be able to articulate knowledge. This can be a cause to, but also an effect of poor verbal or written communication skills.

Knowledge is tacit (ID:47): If the possessed knowledge is tacit, a knowledge bearer will not realize that he has knowledge relevant to others.

Impaired communication (ID:4): If communication is impaired, meaning less frequent and less face-to-face meetings, members of one team will not realize if members of the other team may need their knowledge. This is a common problem in distributed projects [121]. Sub- causes of impaired communication are:

  Asynchronous communication (B): Asynchronous communication (e.g. email or tele- phone) leads to less frequent and thus impaired communication.

  Physical distribution (B): In case of physical distribution of teams, communication be- comes less frequent and thus impaired.

  Different time zones (B): In globally distributed projects, teams may be located at differ- ent time zones, which further limits communication.

Content Perceived Not Useful (ID:18)

If a project participant does not perceive the content of the experience base as useful for his work or current problem, he will not look into the experience base again. These barriers can be found in local, as well as in distributed and collaborative projects. It is important to note that the content in the base may objectively and theoretically (e.g. from the point of view of the knowledge managers) be highly useful for the project participants. However, if the project participant does not *perceive* it as such, he will not contribute or use the base. Content is also perceived as not useful, if team members have no access to the system [179], which was already mentioned above. The barriers that cause the content of an experience base to be perceived as not useful are:

KB scope too broad and shallow (ID:20): If the topic covered by the knowledge base (KB) is too broad, it is very likely that it will be too shallow for project participants to be of value.

Knowledge becomes outdated too quickly (ID:21): Some knowledge items are short-lived and become outdated very soon. If the experience base is not maintained regularly and contains wrong and outdated knowledge, project participants could rightfully question the quality and thus the helpfulness of the bases content.

Experiences are ambiguous (ID:23): An experience is ambiguous, if the solutions it presents may work in one situation and not work in another due to, e.g. a wrong granularity of the experience (see Section 5.3.3) or lack of context.

Content quality is low (ID:19): If the (perceived) quality of the base's content is low, it will not be perceived as useful. A sub-cause is:

> *Lack of knowledge distillation (engineering) (ID:24):* If collected experiences are not engineered, the experience base will not be useful, as raw experiences are hardly reusable [216].

Information overload (ID:62): If the participants are overwhelmed by too much information presented to them in the base, they will not be able to quickly find what they need and consider the experience base useless.

Too little seed (ID:22): If the base contains only little (initial) content, named *seed* [96], it would not be attractive to the project participants to look into or to contribute and they will unlikely find information that helps them.

"Not-invented-here" syndrome (ID:43): This philosophical stance means that creating new knowledge is more prestigious than to reuse existing, and disregard all knowledge that was not "invented" in the company. The collected knowledge will be considered useless.

Knowledge unstructured (ID:59): If the presented and disseminated knowledge is not well structured so that project participants can quickly and easily find the answer to their problem, they will not consider the base useful. Sub-causes, but possibly also effects (circular relations possible) of this cause are:

No context in knowledge / document (ID:61): If the pieces of knowledge documents that should be reused do not contain enough context on how, when and in which situation to apply, they become useless.

Not searchable (ID:60): If the knowledge that is stored in the base is not searchable, project participants will not be able to find the specific information they need making the base useless to them.

No Immediate Benefit (ID:17)

Immediate benefit (also see Section 5.5.1) or short-term benefit is the value that the base holds for the project participants. If they do not see an immediate benefit shortly after the knowledge management initiative has officially started, they will not use the base and will not contribute. The cause is:

No reward / acknowledgement (ID:40): If the project participants, who contribute their knowledge for the public good do not feel rewarded, they will not contribute again. A reward can be monetary, an acknowledgement from colleagues or superiors encouraging preferably an intrinsic motivation [207].

Effort > Benefit (ID:52)

If the process of sharing knowledge the way the EM initiative demands costs project participants more perceived effort and time than the perceived benefit they receive, they will not contribute. This is often the case if a "disparity between who does the work and who gets the benefit" [109] occurs. This disparity can occur when knowledge management tasks serve for the benefit of the collective and not for the individual. The person has to spend considerable time and effort to interact with the system and verbalize (or rank or document) his knowledge and experience. This results in a lack of perceived value and dwindling interest in the EM system [235].

The general lack of perceived benefit (and thus the prevalence of effort) is directly influenced by the lack of immediate benefit. If project participants do not immediately, i.e. from the first use of the system, perceive a personal benefit for their work, they will consider it unbeneficial in the broader sense. So any effort will be considered as too much. The barriers that cause the participants to perceive a knowledge management initiative as causing more effort than providing benefit are:

Instability, system not reliable (ID:57): If the system is not stable in operation and often crashes or loses connection, knowledge bearer will not risk spending time and effort to share their knowledge in vain.

Knowledge management system (KMS) too administratively demanding (ID:54): If the knowledge takes too much effort to administrate, it will be considered as not worth the benefit.

KMS too time consuming to learn and use (ID:56): This more general issue means that if to become acquainted with the KMS and learn to use it takes too much time, the effort will overweight the benefit. The sub-causes are:

   KMS not well integrated in work process and accepted tooling (ID:50): If the implemented experience life-cycle of the EM initiative is not well integrated into common work processes and tooling environment, it will mean extra effort for the project participants to adapt to the new knowledge management-related processes and tools. If they are too unaccustomed with the processes and tools, effort to use them may exceed the perceived benefit. The sub-causes are:

   KMS too technology based (ID:55): If an experience base highly depends on a technology that has to be mastered by the contributors, they will consider it as too much effort.

   KMS has bad usability (ID:53): A bad usability of a system generally means that the user cannot fulfill a task effectively, efficiently and to his satisfaction using this system [3]. The lack of the three criteria leads to a high perceived benefit.

Too time consuming to verbalize knowledge (ID:47): If knowledge is too tacit, it takes too much effort to formulate it in an understandable way. The sub-causes are:

   Shared language not mastered well enough (ID:14): If participants of collaborating teams do not speak a common language well, verbalizing knowledge will be a great effort to them.

   Poor verbal / written communication skill (ID:12): If participants cannot articulate their knowledge well to others, it will be a great effort for them to try.

   Knowledge is too tacit (ID:47): Knowledge that is tacit makes its externalization effortful.

Dispersed teams have their own sharing environments (ID:15): This issue specially appears in distributed collaborative projects. If each team has its own knowledge management systems and processes, it would mean extra effort for project participants to establish a common process. A sub-cause, but also possibly an effect of this cause is:

   Different knowledge sharing culture (ID:3): This issue is similar to the different sharing environments but considers the differences in (organizational) sharing cultures of each team. Adapting to new rules and ways how, where and what to share costs time and effort.

Lack of Trust (ID:9)

Lack of trust is a knowledge sharing barrier that appears in more severe forms in globally distributed projects. There are several definitions of trust summarized by Bertino et al. [37]. *Trust in a system* (not between human parties) is defined as "a belief that is influenced by the individual's opinion about certain critical system features." [150]. A more restricted notion of trust employed by Bertino et al. in a system refers to security management and covers "problems include formulating security policies and security credentials, determining whether particular sets of credentials satisfy the relevant policies, and deferring trust to third parties" [37]. *Trust between business partners* is defined as "the property of a business relationship, such that reliance can be placed on the business partners and the business transactions developed with them" [37], which covers "reliability, dependability, honesty, truthfulness, security, competence, and timeliness" [37].

Both notions of trust are important and the lack of it can lead to not sharing knowledge. If project participants do not trust a system because it lacks secure protocols and no credentials are exchanged between sharing organizational partners, they will not contribute out of NDA policies. If they do not trust colleagues from partner teams to be honest, secure and competent, they will also hoard their knowledge.

Besides the two definitions of trust, there are two types of trust: *cognitive trust* and *affective trust* [136, 127]. Cognitive trust refers to trust in co-worker's reliability and competence. Affective trust is the trusted person's ability to meet emotional expectations that are based on mutual concern and care between the co-workers.

A distributed projects situation can amplify trust problems as people do not often meet face-to-face and do not get to know each other. According to Holste [127], distributed teams "need more time to collaborate efficiently in order to establish trust", otherwise team members will work isolated and not share knowledge. Barriers that lead to lack of trust or magnify it, are:

Knowledge confidence unclear: do not trust others' knowledge (ID:10): If a team member cannot assess the knowledge bearer's competence, he is not certain about the quality and the integrity of the shared piece of knowledge and does not trust this person and his knowledge. Thus, the team member will also not share his knowledge with the knowledge bearer.

Unclear about access to KB after project completion (ID:8): This is often the case, knowledge base (KB) is external and hosted at one collaboration partner's location. Collaboration partners become uncertain and distrustful (especially if they have a competitive relationship outside the project), if they will still receive access to the mutually collected knowledge after the project's end. Thus, members of other teams may either hoard their knowledge or use an own knowledge base.

Knowledge hostile environment (ID:6): A working environment is knowledge hostile, if individual knowledge-sharing behavior is reserved. If people hoard their knowledge, reject external knowledge and have a negative attitude towards mistakes, then a hostility towards knowledge sharing occurs. A sub-cause, but also possibly an effect (circular relationship possible) for knowledge-hostile attitude is:

> *Mistakes are tabooed and not recorded (ID:32):* If the organization, i.e. management, punishes mistakes by e.g. demotion or criticism, team members will not willingly share their negative experiences with others to avoid punishment or degradation. This can lead to a knowledge-hostile environment or a hostile environment, where mistakes are tabooed.

Fear of knowledge parasites ("free-ride") (ID:39): Knowledge parasites are those team members, who do not share but only consume and profit from the knowledge of others. If team members shared a piece of knowledge and have put much effort into acquiring it, they may not be willing to grant the "free-riders" this knowledge.

Poor verbal / written communication skills (ID:12): Poor communication skills hamper knowledge sharing and communication that would make the team member get to know each other better and promote trust. Sub-causes are:

> *Shared language not mastered well enough (ID:14):* If a common language between collaboration teams is not spoken well enough to ensure enough communication, they will not get to know each other well to trust.

> *Different national culture and ethnic background (ID:13):* A different cultural and ethnic background can alienate the teams further and hamper communication and trust between them.

Fear of misinterpretation or criticism (ID:41): As an affective lack of trust, team members may be afraid that their shared knowledge (especially mistakes) will be misunderstood and thus laughed at or criticized.

Protection of power (ID:37): Sharing knowledge can be considered as losing a competitive advantage (bonuses, promotion, protection), power and control. A sub-cause is:

> *Fear to become replaceable (ID:37):* In some companies making mistakes can mean to lose the job or be demoted. People may not trust the company, team members and management fearing repercussions from them if they share mistakes.

30

Do Not Know Where and How to Submit Knowledge (ID:11)

Bad communication flow, especially in distributed teams can mean that team members may not learn how and where to access the knowledge management system and where to contribute their knowledge and experience. Causes, especially inherent to distributed projects, for this lack of awareness are:

Lack of social network (ID:33): Due to lack of knowledge about who knows what and lack of possibility to ask someone, team members may not know where and how to submit knowledge.

Impaired communication (ID:4): If the communication is hampered, it is more likely that information about knowledge sharing possibility will not reach some team members. Reasons for infrequent communication are:

Asynchronous communication (B): Asynchronous communication like email or telephone leads to less frequent communication and information sharing.

Physical distribution (B): Teams that work at different locations have inefficient ways to communicate.

Different time zones (B): Different time zones further limit the communication possibility.

3.3. Discussion

Each knowledge sharing barrier listed in Tables A.1 and A.2 is a direct result of the literature review. Figure 3.1 depicts a subset of the barriers with causal relationships. Many of the causal relationships and associations of sub-causes, causes and main causes could only be indirectly derived from notions in the literature. Due to a cyclic nature of some barrier relationships, my overview structure is one of many possibilities. To prove or refute the causal relations is elaborate: each relation must be examined in a controlled environment negating any outside influences. This could be approached with the help of qualitative methods [81].

The choice of the barrier titles may also be debatable. Many barriers do not have an unanimous naming in the literature. I had to decide, which barriers to consolidate as the same and which to differentiate. This choice may be subjective. In some cases, barriers I considered synonymously named may have a slightly different connotation. This is also a decision of granularity. Aspects could be considered separately as more fine-grained causes or consequences. I made a decision to consider them as the same, if their granularity was so fine, that they would distract from to the main research question of this thesis.

31

Many of the reviewed sources group their barriers according to categories like personal issues, management issues, cultural issues or technical issues. This is similar to my structure. I, however, identified cause and effect relationships between many barriers. In the literature most of the cause and effect relationships were explicitly or implicitly mentioned. On some interconnections I decided through implicit information in the literature.

The goal of the literature review was to collect barriers to experience and knowledge sharing in the distributed environment. The decision which barriers to take into the final overview was not trivial. Many found barriers are fundamental and can be considered for distributed joint ventures as well as co-located and company-wide projects (e.g. bad usability). In my attempt to sort out only those barriers that are especially apparent in distributed software engineering, I also had to make a choice of fundamental barriers in all categories, which directly influence the main barriers.

The methodology of this review lacks several aspects of a systematic literature review as suggested by Kitchenham et al. [151]. I did not explicitly protocol which source was retrieved from which database. I also lack a protocol about which literature directly or indirectly mentions a causal relationship between barriers.

3.4. Contributions

This chapter answers research question **RQ1**: *What are significant experience sharing challenges in distributed collaborative software development projects?*

A key insight from this chapter for the argumentation of this thesis is that most of the knowledge sharing barrier categories directly affect the effort / benefit ratio. It also showed that lack of time is the most frequently mentioned knowledge sharing barrier in the reviewed literature. The review also revealed that without access to knowledge, developers will not see any (immediate) usefulness in the base's content and will not be motivated to spend time sharing their experience. Without enough perceived (immediate) personal benefit, they will not make an effort to share. Even lack of trust can partially be a cause for increased effort. Not trusting project partners or having non-disclosure policies means that experience bearers will have to spend additional effort on deciding which information is allowed to be shared with whom. Thus, effort and benefit imbalance and lack of time can be identified as a major sharing inhibitors beside lack of awareness and trust.

criteria in Definition 4.3. If none of the criteria are fulfilled, then I classify a method as heavy-weight. To assess if a method is more light-weight than another or that the method *is* light-weight, all criteria have to be compared only to methods of the same method category or with methods having the same goal for experience collection.

Definition 4.3 (Light-weight criteria).

An experience collection method is light-weight, if it conforms to or strives for at least one of the following criteria (C):

C1 Software that the experience bearer is acquainted with is reused or extended.

C2 Collection is well integrated in the experience bearer's existing working process(es) and environment.

C3 The time to share an experience is kept short.

C4 Cognitive load is reduced or minimized making experience sharing easy to perform for the experience bearer.

C5 Experience engineering is shifted away from the experience bearer to others.

Criterion C1 sets a requirement to the tool(s), which the experience bearer explicitly uses to share experiences or initialize the sharing process. C2 is concerned with the collection process and knowledge sources that are encoded into organizational routines [104]. Often C2 results in C1 or the other way around. Nevertheless, C2 can be fulfilled without C1, if the method tries not to involve the experience bearer in the collection process by exploiting certain events or analyzing artifacts that may contain experience. C3 and C5 directly aim at lowering the sharing effort and invested time. C4 often results from C1 and C2, but can also be a standalone distinguishing criterion.

The progression from a light-weight method to heavy-weight is a continuous and fuzzy scale. Based on Definition 4.3, I assume a threshold: An experience collection method is light-weight, if it complies to at least one criterion. I also assume that a method complying with all five criteria deeply will unburden the experience bearer more than a method just at the threshold. However, Definition 4.3 does not provide an assessment of light-weightedness related to the *extent* of a fulfilled criterion. It only provides a (ordinal) scale to compare the light-weightedness according to the *number* of fulfilled criteria. Thus, it is not possible to decide, if a collection method A that only marginally fulfills all criteria from Definition 4.3 is more light-weight than a method B that only fulfills a few criteria, but in a much greater extent. In this case, there is a need for a measurement system, which is approached in Chapter 5.

If *no* criteria are fulfilled, I define the method as heavy-weight. It is important to note that though the light-weight and heavy-weight method sets are only theoretically disjoint: The decision to sort a method in one or the other set is often difficult and can be subjective. It is often hard to determine, if the method *fulfills* a criterion from Definition 4.3 and if it fulfills it *enough*.

4.2. Lowered Experience and Knowledge Sharing Barriers

This section analyzes which knowledge sharing barriers from Chapter 3 can be lowered by light-weight knowledge collection approaches.

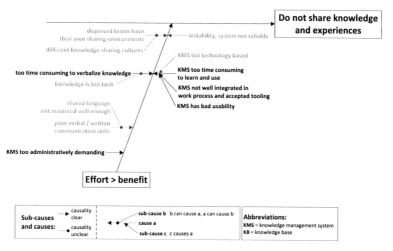

Figure 4.2.: Excerpt from the knowledge sharing barriers overview (Figure 3.1) depicting those barriers (black) that are directly overcome by light-weight experience collection methods. Other barriers are grayed out.

Figure 4.2 highlights (by black color) the sharing barriers that are overcome directly by the light-weight method concept. These issues are the basic set, meaning that if a knowledge collection method is light-weight, it always tries to lower at least this set of barriers. Other barriers that are not overcome by light-weight methods are grayed out. Concrete light-weight method instantiations, which are presented in Section 4.3, may overcome more issues than the highlighted ones. The light-weight concept is independent from the project's distribution. Thus, it does not directly resolve issues that are special to globally distributed collaborative projects. Knowledge sharing methods and environments that also consider distributed software development, are presented in Chapter 6.

According to Definition 4.1, the main aim of a light-weight collection method is to lower the effort for the experience bearer in his task of knowledge sharing. Thus, the main focus of the light-weight concept is to leverage the *effort > benefit* issue, which points out a disparity between too much effort to share experience and too little benefit from the experience management initiative.

More specifically, criteria C1 (reuse known software) and C2 (integrate into process and tools) in Definition 4.1 suggest to reuse familiar software and not alter the usual workflow and environment. C2 directly resolves the issue *KMS not well integrated in work process and accepted tooling* and C1 *KMS too time consuming to learn and use*. C3 (keep sharing process short) prescribes to minimize the experience bearer's time to share, which also supports the latter issue. Abiding by C3 should also help the experience bearer to either lower the *time to verbalize his knowledge* or relieve him of verbalizing it and choose other ways to capture his experience. The barrier of not using the knowledge management system because of *bad usability* can be lowered by minimizing the experience bearer's cognitive load (C4), integrating the system into the experience bearer's working environment and making the process of learning and sharing knowledge as short as possible (C3). These are the main goals of usability engineering. Fulfilling C2 means to raise learnability, C3 and C4 would raise the efficiency of a software [189, pp. 26-37]. Criterion C5 prescribes to shift experience engineering task to other people than the experience bearers. This would solve the issue of a *too administratively demanding knowledge management system* by delegating these tasks to experience engineers.

4.3. Experience Collection Method Categories

In order to be able to quantify light-weightedness of experience collection methods, I conducted a literature review and identified collection methods that abide by Definition 4.3. They are presented in Section 4.4. The listing of these methods can also serve as a catalogue for a knowledge manager. In this section, I describe the method categories the light-weight methods belong to. This way, the knowledge manager can choose a method of the appropriate category that suits the specific situation. Table 4.1 presents an overview of the various experience collection method categories and sub-categories.

The overview of the experience collection methods in Table 4.1 is based on Dalkir [74] and Schneider [216, p.144]. This overview of the experience collection method categories and sub-categories does not claim to be exhaustive. Some literature (e.g. [67, 41, 226]) suggest partly different collection method categories with a more extensive number of methods. However, I

Table 4.1.: Overview of the different experience and knowledge collection method categories with more specific sub-categories. The italic method categories cannot be instantiated as main methods, only as secondary methods (see Section 2.1.2). SE is an abbreviation for software engineering.

Method category	Sub-category	Description
experience as a by-product		Capture experience as a by-product during a SE activity.
data mining	text analysis	Retrieve experience by analyzing SE documents.
interview	fully structured interview	The interviewer strictly follows a script.
	semi-structured interview	The interviewer follows the script more loosely.
	unstructured interview	The interviewer does not prepare any questions.
experience workshop	post-mortem workshop	This meeting to share experiences takes place after the project ends.
	anytime workshop	This meeting to share experiences can take place anytime during the project.
experience authoring		Experience bearer externalizes his experience by e.g. writing down.
reflection-in-action		Creating knowledge and reflecting on it while working.
collaboration support		This method is constructed to support collaboration tasks and processes.

consider them too specific or not suitable in the scope of this thesis, as they are designed to collect *knowledge* or even *information* and not *experience*. I chose those method categories and sub-categories in the literature, which were widely mentioned and for which I identified at least one light-weight method instantiation.

Next, each method and its sub-methods are introduced in more detail. For each method, the knowledge conversion (according to Section 2.1.2) is described. Additionally, for each method a graphical illustration in FLOW notation (see Section 2.1.4) in the SE domain is given.

The methods emphasized in italic font cannot be instantiated as main methods, only as secondary methods.

4.3.1. Experience as a By-Product

Experience as a by-product (also further abbreviated as "by-product method") is a method to both elicit and extract knowledge and experience with the main goal to shift effort away from experience bearers and from the time the knowledge containing tasks are carried out [215]. The *rationale as a by-product* approach to capture decision rationale was described by Schneider [215]. The approach is described by two goals and seven principles. In the context of this thesis, I reuse Schneider's definition, but transfer it from rationale to experiences. Hence, experience as a by-product is characterized by the following goals and principles:

Goals

1. Capture experience during specific tasks within software projects.

2. Be as little intrusive as possible to the experience bearer.

Principles

1. Focus on a project task in which experience is surfacing.

2. Capture experience *during* that task (not as a separate activity).

3. Put as little extra burden as possible on the bearer of the experience (but maybe on other people).

4. Focus on collecting experience only during the original activity. Defer engineering to a follow-up activity carried out by others.

5. Analyze recordings or text, search for patterns.

6. Encourage, but do not insist on further experience management.

Figure 4.3 illustrates the processes, general roles, as well as information and experience flows in the experience as a by-product method. It shows the most important characteristic of the method: software engineers (experience bearers) only do their regular jobs and are not explicitly involved in experience sharing. Experience is then extracted *during* the software development process.

It is important to differentiate between this approach and data mining in Section 4.3.2, as both method categories have overlapping defining characteristics. It can be argued that data mining also discovers knowledge and experience contained in documents as "by-product", i.e. without (almost) any effort from the document author. In this thesis, the *by-product* concept is defined as capturing knowledge and experience from a source *during* its creation or uttering and not afterwards, as do data mining approaches. As an example, recording a prototype meeting and

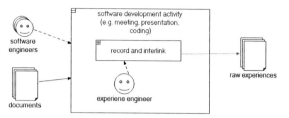

software engineers

software development activity
(e.g. meeting, presentation, coding)

record and interlink

raw experiences

documents

experiene engineer

Figure 4.3.: Main roles, activities, information and experience flows for *experience as a by-product*.

linking it with the presented code and slides (see the *FOCUS* method in Section 4.4.3) would be a by-product method [215]. Using video or audio recognition algorithms on the already recorded video to extract relevant segments, would be data mining.

Concerning knowledge conversion, this method is used to externalize implicit knowledge into explicit recording and combine it by interlinking with other explicit knowledge (code).

4.3.2. Data Mining

Data mining (Figure 4.4) is a method to automatically extract or discover "predictive information from large data bases" [74]. This is often achieved by statistical analysis, database techniques

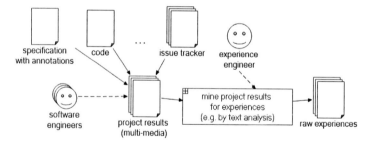

specification with annotations

code

issue tracker

experience engineer

software engineers

project results (multi-media)

mine project results for experiences (e.g. by text analysis)

raw experiences

Figure 4.4.: Main roles, activities, information and experience flows for *data mining*.

and clustering. The goal of data mining is to search for patterns hidden in the data and explain them according to models and rules. Data can be numbers, text, graphics, audio or video stream, etc. Possible representations of knowledge uncovered by data mining could be e.g. result of a statistical analysis, an ontology (see Gruber for definition [107]), a string of text denoting an identified trend or cluster in the mined data.

The quality of results, usually depends on the number of input data (the more the better) and the types of models. Simple models are e.g. regular expressions or decision trees, while more complicated techniques like neural networks mostly remain "black boxes", i.e. non-transparent to users [74]. For further reading on the wide research field of data mining be referred to e.g. Han et al. [111] or Witten et al. [257].

Text analysis: In this thesis, text analysis is considered as a special case of data mining operating specifically on text documents written in natural language. It is also part of the field of natural language processing (NLP), an "area of research and application area that explores how computers can be used to understand and manipulate natural language text or speech" [59]. Here, text analysis also includes manual processing of text to uncover patterns.

As data mining methods work on externalized data, e.g. text, they can only extract explicit knowledge (though they may find new relationships that are implicitly hidden in the data) and combine it to another explicit knowledge representation.

4.3.3. Interview

An interview is a method to elicit information from people as source. It is "more or less a structured conversation" between two people or a group [216, p.166]. Interviews can have different question types – open and closed – with different goals in mind. *Closed* questions provide predefined and limited set of answers (often as a word, value or sentence) to a question, while *open* questions allow a broad range of answers and result in answers that are usually longer than a word or sentence [205]. Closed questions aim at "discover[ing] the responses that individuals give spontaneously" [223], while open questions are useful "to avoid the bias that may result from suggesting responses to individuals" [223]. Open questions demand usually more time and effort from the interviewee than closed questions [216, p.167]. Filling out a questionnaire is also considered as an interview here, as the process is the same – questions are posed and answered, only without a human interviewer. There are mainly three types of interviews divided by their structure [198].

Fully structured interview: A fully structured interview follows a strict sequence of questions.

Semi-structured interview: In this kind of interview, questions are prepared but followed more loosely. The interviewee talks more freely, than in a structured interview, as it contains more open questions.

Unstructured interview: For this interview, the interviewer does not prepare any questions.

Figure 4.5 presents the two variants how an interview can be conducted. The left FLOW diagram depicts the usual interview process by a human interviewer, while the right diagram shows the special situation where the software engineer fills out a questionnaire and not interviewed led by a human. For more details on interviews about different processes and types be referred, e.g., to Runeson and Höst [205] or Shadbolt and Burton [226].

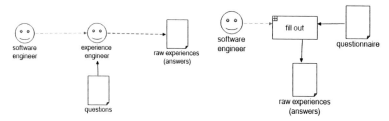

Figure 4.5.: Main roles, activities, information and experience flows for *interview*.

From the knowledge conversion point of view, interviews help to externalize knowledge from tacit and implicit in peoples' head to an explicit protocol or recording. In more loosely structured interviews with open questions, tacit knowledge is more likely to be elicited than in a fully structured interview with closed questions.

4.3.4. Experience Workshop

Experience workshop is a sub-method of the more general workshop in the context of this thesis. The specialization lies in the topic of the workshop. An experience workshop has always the main goal to elicit experiences.

An experience workshop (Figure 4.6) is a meeting that serves the goal to elicit experiences for the benefit of future projects [216, p.169]. Unlike interviews, experience workshops serve to elicit knowledge not from one but from a group of people. In contrast to an interview, each participant does not get as much personal attention as it is divided by more participants. On the other hand, participants can interact and directly share knowledge with other participants. Unplanned reactions from other participants to a story can activate tacit knowledge. An experience workshop should be planned and have a structured process [216, p. 169]. Usually, the results of the workshop have to be documented "in a form appropriate for later reuse" [216, p.169], e.g. in a report, by the workshop organizers.

There are many workshop variants. This thesis differentiates between the time the workshops are conducted [216, p.167].

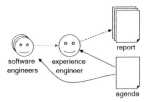

Figure 4.6.: Main roles, activities, information and experience flows for *experience workshop*.

Post-mortem workshop: A post-mortem workshop or review or also called retrospective is held (usually once) at the end of a project. The goal is to elicit project experiences after the end of the project [216, p.169].

Anytime workshop: This type of experience workshop can be held repeatedly and anytime during the project.

For knowledge conversion, workshops can elicit implicit, but also tacit knowledge and externalize it into explicit documents or recordings.

4.3.5. Experience Authoring

Experience authoring (Figure 4.7) is defined here as a more specific method than the general task of content creation. In this thesis, experience authoring is the *explicit*, i.e. conscious, writing down of one or more pieces of experience by the bearer of the experience [216, p.144]. A very simple way, for example, is to take a pen and write the experience down on a sheet of paper.

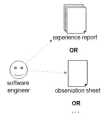

Figure 4.7.: Main roles, activities, information and experience flows for *experience authoring*.

Techniques of this method category support the experience bearer in writing experiences. They can be tools that make him spend less time on the task or make him aware which of his experiences are valuable and provide help on how to write them down.

Written experience can be of various size, ranging from a couple of sentences [230, 23] to e.g. a 100 page project post-mortem experience report [71]. Specific tasks like writing a blog or annotating text are also considered experience authoring here. Blogs fulfill the aspect of writing down experiences in form of emotional stories [187]. Annotations are small bits of explicitly written knowledge and possibly experience.

Experience authoring has a definition overlap with the interview method (see Section 4.3.3). Filling out a questionnaire or survey is considered a special case of interview here, as the question and answer pattern is fulfilled, only they are not conveyed orally but in written form. Writing down answers (experiences) also fulfills the definition of experience authoring. This special case is defined as a composite method with interview as the main and authoring as the secondary method.

Concerning knowledge conversion, experience authoring provides the possibility to externalize implicit knowledge. Combined with other methods, e.g. reflection-in-action (see next section), also tacit knowledge can be activated and elicited.

4.3.6. Reflection-In-Action

Reflection-in-action is a theory by Donald Schön [221] on how "working and creating knowledge interact" [216, p.39]. He observed that professionals *while* working, hardly reflect on *what* they are doing. Afterwards they cannot explain what they did because they worked in "tacit mode" and did not remember the details anymore. Schön states that interruption, also called a "breakdown", of the experience bearer during his work, helps him to reflect. Breakdowns lead to reflection (in action) about the experience bearer's doing, making him aware of his tacit knowledge. By this, tacit knowledge becomes implicit and can be elicited. The *method* of reflection-in-action is thus to induce breakdowns during a work activity.

An experience collection approach falls into this category, if one of its main intentions or design is to make the knowledge worker reflect by interrupting him during his task, most often through a feedback component of some kind. A feedback component is a method that can initiate a breakdown. *A critiquing system* introduced by Gerhard Fischer [98, 95] is a prominent example that uses the reflection-in-action paradigm. A critique is "a system that presents a reasoned opinion about a product or action generated by a human" [97].

Reflection-in-action can only be instantiated as a *secondary method*. It is not a standalone method, as breakdowns always require a task. This task would be the main method. For example, if reflection-in-action occurs through an unexpected question in a discussion, the main method would be an experience workshop or interview. In case of a feedback component in an experience writing tool, the main method would be experience authoring.

The roles, information flows and activities of reflection-in-action are depicted in Figure 4.8.

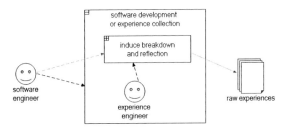

Figure 4.8.: Roles, activities, information and experience flows for *refleciton-in-action*.

In the diagram, a software engineer either conducts a development activity or takes part in an experience collection activity. During this activity a tool or method managed by the experience engineer creates a breakdown and induces reflection. Experiences are activated by reflection, which is depicted by the experience flow from the software engineer into the reflective activity. As an output from this activity, these experiences are externalized as raw experience documents. Reflection-in-action does not define how exactly experiences are externalized and who creates the documents.

Concerning knowledge conversion, reflection-in-action methods can achieve externalization or socialization of implicit and tacit knowledge to implicit or explicit.

4.3.7. Collaboration Support

This method category aggregates those methods that have the goal to support distributed communication and collaboration that can contain knowledge. Often, but not exclusively, these methods include massaging systems, email, electronic meeting, workflow support or Web-based applications supporting collaboration [74]. Communication can be synchronous or asynchronous. Writing an email, for example, is asynchronous, while writing through a chat messenger or video conferencing supports synchronous communication [21].

Like reflection-in-action, collaboration support can only be instantiated as a secondary method. There are many method instantiations which mainly aim at facilitating collaboration and consider knowledge exchange a sub-task. In this thesis, though, the focus lies on experience collection. Collaboration support is a method that does not directly collect experience. This method category state that an instantiation is suitable to be employed in a distributed collaborative project. Methods that do not instantiate collaborative support may also be employed in a distributed project, but may need additional tooling or processes.

Figure 4.9 shows distributed teams that participate in the activity of experience sharing. This is

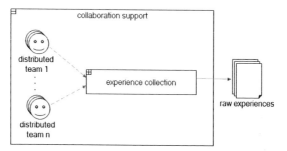

Figure 4.9.: Roles, activities, information and experience flows for *collaboration support*.

done within the collaboration support method (in the diagram within the activity). Collaboration support facilitates distributed experience sharing.

From the knowledge conversion point of view, any conversion can take place during collaboration. Video conferencing, for example, makes the participants socialize and transfer tacit and implicit knowledge. Recording or transcribing it externalizes this knowledge, as well as a chat. Asynchronous tools like email are a source that can be used for combination of explicit to explicit knowledge.

4.4. Light-weight Method Instantiations

This section provides an overview of the state-of-the-art on *light-weight* experience collection. In the following section, I first describe the methodology of the literature review, followed by the results. There, I list the method instantiations according to the method categories (as introduced in the previous sections) and argue why each method instantiation belongs to the particular category and why it is light-weight. For the latter I investigate, which of the five criteria from Definition 4.3 on page 35 are fulfilled.

4.4.1. Review Methodology

The literature review on light-weight experience collection methods consisted of a database search with a search string, complemented with a recursive search within the found literature. The search string consisted of parts displayed in Table 4.2.

The overall search strings were the elements of the power set of the key words in Table 4.2. Since the community on experience collection is not excessively big, I considered a generic

Table 4.2.: Keywords for the search.

| A1—knowledge | B1—light-weight | C1—collection |
| A2—experience | B2—lightweight | C2—elicitation |

search on Google Scholar and the Emerald database as sufficient, where I expected to contain the most literature on knowledge management (e.g. the Journal of Knowledge Management). I define the following inclusion criteria in the choice of literature:

- The method explicitly mentions or describes *knowledge or experience collection*. I did not limit my search to experience. The reason is that many methods do not explicitly collect experience but more generally knowledge. From the method description it is however often clear that they also include experience in the collection. My focus lay on choosing those methods, where the collected knowledge *may contain* experience. A negative example that was excluded is a method to automatically extract information from meeting minutes into other tools like an issue tracker or Wiki [179]. This method is light-weight in many aspects but it is excluded here as the idea and implementation only focuses on collecting decisions and tasks. At this point, it would be reasonable to argue that decisions and tasks base on rationale, which is often based on experience and which is also often included in a meeting minute. However, this method and tool support were specifically designed to extract solely decisions and tasks without the rationale behind them. Besides, I consider capturing rationale behind decisions and tasks a separate collection method. Examples would be the method instantiations *FOCUS* [215] or *Automatic Rationale Extraction* [199] presented in the next section.

- I did not limit my search to the software engineering (SE) or distributed software engineering (DSE) domain considering that an experience collection method can often be reapplied to other domains. The reason for not limiting on SE and DSE is that a software project also contains activities that are not SE-specific. Examples are a (general) meeting or a post-mortem workshop[1]. These activities can be conducted in a project of any domain. Conducted in a SE project, both activities can be used to elicit SE-related experiences. However, I excluded methods, which core idea is (non-SE or non-DSE) domain specific and cannot be reapplied to SE or DSE without losing the core concepts that define the method.

- I included those methods, where the literature explicitly states that the method conforms to or strives for at least one criterion from Definition 4.3. Otherwise, the method was not

[1]It is worth to note that a meeting is an activity that is part of a project, whereas a post-mortem workshop is a method specifically for experience elicitation.

considered light-weight. However, few sources of literature *explicitly* describe features that can give a clue if the method conforms with Definition 4.3 and is light-weight. If it is not explicitly mentioned, it must be obvious from the method description or application that one or more criteria from the definition are fulfilled. For example, consider the method *Automatic Rationale Extraction* [199] (presented in the next section) that automatically extracts rationale from a document. I classify it as light-weight. Being an automatic extraction process, it obviously unburdens the experience bearer and fulfills criterion C3 by not involving him in the experience collection process.

- I excluded those experience collection methods that do not fulfill any of the criteria from Definition 4.3, even if they are (directly or indirectly) referenced as "light-weight" or "lightweight" in the literature.

4.4.2. Results

As a first overview, Figure 4.10 relates the general collection method categories to the identified light-weight instantiations. The method categories are displayed in gray boxes. The specific method instantiations (white boxes) are placed inside the respective swim lanes of the *main method* they instantiate. In Figure 4.10, the secondary methods (see domain model in Section 2.1.2) are indicated by the placement of the small black circles that are connected to the instantiation by a dashed line. A method instantiation can instantiate more than one secondary method. The method categories with italic font can only be instantiated as secondary methods.

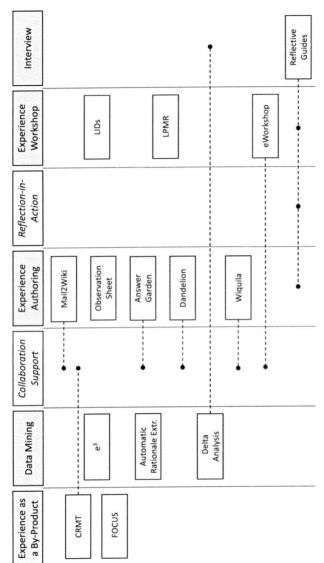

Figure 4.10.: Overview of all experience collection method instantiations and their method classification.

Table 4.3.: Overview of the light-weight experience collection method instantiations. They are grouped by their main method category, including the light-weight criteria they fulfill according to Definition 4.3 and a short synopsis.

Category	Method	Light-weight aspects	Synopsis
experience as a by-product	Collaborative Risk Management Tool (CRMT) [215]	2 - 5	Captures risk assessment rationale as distributed chat and interlink statements to risk placement in portfolio.
	FOCUS [211, 215]	1 - 3, 5	Records prototype rationale from prototype presentation (slides, speech) and link to executed prototype methods.
data mining	Email Expertise Extraction (e^3) [51]	1 - 3, 5	Builds an expertise graph of employees by analyzing content and communication patterns of emails.
	Automatic Rationale Extraction [199]	2, 3, 5	Algorithms identify and extract rationale in project documents.
	Delta Analysis [214]	2, 3, 5	Analyzes closely related project documents for differences. Differences can point to experiences.
interview	Reflective Guides [170]	2, 4, 5	Participants fill out a questionnaire during a task making them reflect on the task.
experience workshop	eWorkshop [32, 33]	3, 5	A tool and process for distributed chat-based workshop to capture experience.
	LIDs [212]	3, 5	The workshop method collects experiences from participants in a protocol and links other documents to places in the protocol they were referenced.
	Light-weight Post-mortem Review (LPMR)[87]	3, 5	A short version ($\frac{1}{2}$ day) of a post-mortem workshop to collect project experiences.
experience authoring	Answer Garden [10]	3	A method and tool to relieve an expert of answering too many (repeating) questions by asking the seeker diagnostic questions and possible answers first.
	Dandelion [58, 57]	1, 2, 4	This is a Wiki plugin to facilitate collaborative authoring of experiences in a Wiki using an email client as authoring tool.
	Observation Sheet [230, 213, 148]	1 - 5	A paper form to write down a short ad hoc experience.
	Mail2Wiki [113]	1 - 5	Three interfaces facilitating to contribute to a shared Wiki using email.
	Wiquila [202]	2, 3	A method that facilitates experience authoring in a Wiki providing context-sensitive link suggestion and a WYSIWYG editor.

Table 4.3 sums up the most relevant method instantiations' characteristics displaying them in an alphabetical order. The table displays which light-weight criteria from Definition 4.3 are covered by the instantiation and presents a short synopsis of each approach.

In the following, each light-weight method is introduced in detail: a description of the idea and the fulfilled light-weight criteria from Definition 4.3.

Concerning the light-weightedness criteria, the overview only mentions those that are fulfilled. For the other criteria, no definite statement is possible due to insufficient information. This is often the case for C2 (integrate into process and tools). It is difficult to evaluate, if a method is well integrated in the work process, if no sufficient information is provided about the process. It is more likely for a proprietary tool to be newly introduced to the experience bearers and thus need learning time. But it would also be dangerous to assume that a common technique like a *shared folder* is well accepted by default: The team might have always used proprietary groupware. It is important to note, that I do not regard an experience management initiative as a whole, but only the collection process.

4.4.3. Experience as a By-Product

Collaborative Risk Management Tool (CRMT)

Description: The goal of the Collaborative Risk Management Tool (CRMT) [215] is to capture rationale behind a risk assessment meeting. Risks are assessed by placing them in a risk portfolio. The rationale is contained in the discussions and explanations around the risk placement. CRMT provides a tool containing a risk portfolio and allows distributed risk assessment. The discussions are captured in form of a chat component that is integrated in CRMT next to the portfolio. The chat is just a sample implementation and can be replaced by any other recording technique. The chat is synchronized to the placement of risks in the portfolio. A risk manager in the next project, for example, will be able to comprehend why a risk was placed at a certain position and then moved by reading the corresponding statements in the chat. Figure 4.11 visualizes the process, communication and documents of CRMT.

Method: CRMT is a composite method. Mainly it falls under the category of experience as a by-product (recording and interlinking), as it fulfills the goals and principles of *experience as a by-product* [215]. As a secondary method, it supports collaboration, as CRMT allows distributed risk assessment meetings.

Light-weight criteria: Criteria C2–C5 are fulfilled. The collection is well integrated into the usual risk assessment process as it does not change it in any way (C2). The software engineers do not perform any additional work (C3 and C4) lowering the time to share and effort and cognitive load. The only effort for them is to use CRMT. According to the definition of a by-product

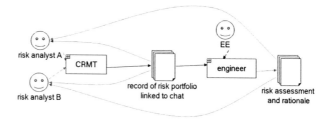

Figure 4.11.: Process, communication and documents with the Collaborative Risk Management Tool.

method (refer to Section 4.3), C5 is also fulfilled.

FOCUS

Description: FOCUS is a "strategy and a family of tools" [215] to capture rationale behind software prototypes and specifically answer questions like "why do it that way?" [215]. These answers can be effectively elicited during prototype demonstration meetings, as "[d]emos often convey highly condensed information, far beyond 'raw rationale'"[215]. Similar to CRMT above, FOCUS captures discussions around a prototype and synchronizes them with the object of discussion. FOCUS records a prototype presentation as screen capture and explanations as audio using e.g. Camtasia[2] and interlinks it with the "sequence of executed methods" [215] in the code. FOCUS is implemented as an Eclipse[3] plugin, to integrate it into the environment that is used for "writing, running and explaining code". The prototype experts have to start the recording and code tracing processes by "press[ing] a few buttons"[215] in the FOCUS panel. Figure 4.12 presents a general overview of the experience flow and documents.

Method: FOCUS follows the by-product approach capturing experiences during a SE project event unburdening the prototype developers from sharing their experience as an extra task.

Light-weight criteria: C1–C3 and C5 are fulfilled. C1 (reuse known software) is fulfilled, as FOCUS is implemented as a plugin for an IDE the developers work with. The experience collection is part of the prototype presentation meeting using the working process of oral presentation and code execution as input (C2). The bearers do not have any additional effort according to the by-product method definition (C3). According to the definition of the by-product approach C5 (shift experience engineering away) is also fulfilled.

---

[2]http://www.techsmith.de/camtasia.html
[3]http://www.eclipse.org/

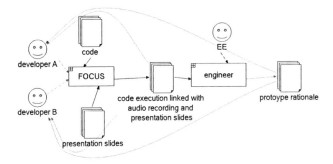

Figure 4.12.: Process, communication and documents with FOCUS.

4.4.4. Data Mining

Delta Analysis

Description: Delta Analysis is a knowledge discovery method that aims at finding "patterns in documents that point to interesting experience" [214]. The method is based on the heuristic that a "difference ('delta') [between documents] could be caused by (1) conscious decision or (2) mistake, both of which need closer attention" [214]. The experience engineer (manually or semi-automatically, if a version control exists and is applicable) compares two closely related documents for differences. Examples of such documents are two subsequent document versions, a graphic and its textual description or a table and its MS Excel sheet representation. After uncovering the differences, the experience engineers have to investigate the reasons why these differences occurred. This is done by reading more documents or conducting focused interviews with document authors. Figure 4.13 presents a general overview of the experience flow and documents.

Method: Delta Analysis belongs to the text analysis method category as the main method to uncover document differences as the main method. Interview method is used as a secondary method if further investigation of document differences is needed.

Light-weight criteria: C2 (integrate into process and tools), C3 (keep sharing process short) and C5 (shift experience engineering away) are fulfilled. The analysis takes advantage of the document creation and editing process of creating new document versions after each change [214] (C2). Since the analysis is supposed to be conducted by the experience engineers [214], C3 and C5 are fulfilled. The interviews, if ever needed, pose the only additional effort the document

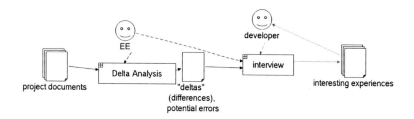

Figure 4.13.: Process, communication and documents with Delta Analysis.

authors have to endure. These interviews, however, are focused (and presumably shorter) than without prior text analysis.

Email Expertise Extraction (e^3)

Description: e^3 is a method to collect knowledge about experts in a project or company [51]. It replaces the usual process of asking colleagues and following referrals to find the right expert for a problem. A suitable source are emails. They can indicate expertise location, since experts often converse their knowledge and experience over email [51]. Emails also can give a notion about who is expert on which topic, "as people routinely communicate what they know" [51]. e^3 presents an algorithm that can discover experts in a particular topic. This is a graph based ranking approach, taking into account both the email content and communication patterns. The algorithm analyzes emails by keywords, clustering them by topic through unsupervised email content analysis and then building a weighted directed graph of senders and receivers. This graph represents the information flow among the people involved. Then, each sender and receiver is given an expertise rating. Technically, the algorithm is deployed on a dedicated server. To include an email in the processing, the sender must add the server's email address as carbon-copy. Figure 4.14 presents a general overview of the experience flow and documents.

Method: The method instantiates text analysis, as it implements an algorithm that processes emails.

Light-weight criteria: C1–C3 and C5 are fulfilled. e^3 reuses the familiar email system as collection mechanism (C1). The experience collection process is rather well integrated into the working process and tooling: The author needs to carbon-copy the email to the e^3 processing server, which does not significantly alter the usual email sending process (C2). The email authors are barely involved in the expertise mining and the analysis afterwards besides adding an

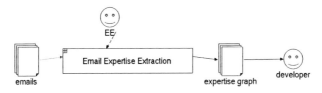

Figure 4.14.: Process, communication and documents with e^3.

additional email address (C3 and C5).

Automatic Rationale Extraction

Description: Decision rationale in software engineering is a valuable asset [90] that should be captured and reused. Attempts to encourage "explicit rationale capture" were unsuccessful, as developers were not willing to spare time for this "tedious, intrusive and potentially expensive" task [199]. Assuming that rationale "is often buried in the many forms of documentation", Rogers [199] conceived two algorithms that mine existing project documents for rationale. The first rationale identification method works with text mining based on an ontology (used from an existing source), while the second method is based on recognition of linguistic features that are special to rationale description. These are modal auxiliaries, adverbial and projective clauses. For further technical details of the algorithms refer to Rogers [199]. Figure 4.15 presents a general overview of the experience flow and documents.

Figure 4.15.: Process, communication and documents with Automatic Rationale Extraction.

Method: The general method category behind this method (and both algorithms) is text analysis.

Light-weight criteria: C3 (keep sharing process short) and C5 (shift experience engineering away) are fulfilled. As motivated, the method's main goal is to relieve the bearers of rationale from explicitly sharing it (C3). From the same motivation, though not explicitly stated, it can

55

also be assumed that the rationale bearers are also relieved from all further experience engineering tasks (C5).

4.4.5. Interview

Reflective Guides

Description: The concept behind reflective guides is a "model for managing the knowledge and experience team members acquire during software development projects" [170]. The idea of this approach (focusing on the collection part of the model) is, compared to post-mortem elicitation methods, to capture experience *during* a task. The reflective guides are questionnaires that "are assigned to the software project team members who, during the time in which they execute project tasks, use these guides to register their reflections and impressions, difficulties found, unforeseen facts and similar considerations related to the manner in which they carry out his/her tasks" [170]. Such a task is, for example, a requirements elicitation session. After the process step (requirements engineering) is done, a workshop with the team members is conducted to analyze, compare and reflect on the different answers in the filled out questionnaires. It is also an opportunity to learn what other colleagues answered in their guides and discuss their solutions. After the workshop, the organizers (assumingly the experience engineers) sum up all questions and answers in a summary document, which is then distributed to the participants. Figure 4.16 presents a general overview of the experience flow and documents.

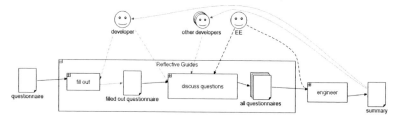

Figure 4.16.: Process, communication and documents with Reflective Guides.

Method: The main method category is a fully structured interview. The main task of the experience bearers is to fill out the questionnaires called reflective guides, which is defined as interview (see Section 4.3.3). According to the definition of experience authoring, filling out a questionnaire also means that this is an experience authoring instantiation. Further, the Reflective Guides implement an experience workshop, which is conducted after filling out the

56

questionnaires. This instantiation has also aspects of reflection-in-action, since the guides are filled out *during* the task, resulting in breakdowns and reflection about the task.

Light-weight criteria: C2 (integrate into process and tools) and C4 (reduce cognitive load) are fulfilled. The collection method conforms to C2 because the authors claim that it works in a "non-disruptive way, by integrating its management into daily project activities" [170]. Non-disruptiveness of the approach also aims at minimizing the cognitive load (C4).

4.4.6. Experience Workshop

eWorkshop

Description: The eWorkshop is a tool-supported process for globally distributed on-line meeting to "synthesize new knowledge" [32] in an SE project. Its goal is to gather distributed experts and elicit their knowledge and experience in "an efficient and inexpensive" way to populate the project experience base. The eWorkshop contains a Web-based chat-application including panels for agenda, message board where the chat messages are displayed, whiteboard for discussion summary and voting, attendee list, chat log and a Frequently Asked Questions (FAQ) directory. The chat containing the knowledge including a timestamp and author are captured to be analyzed later by experience engineers. The eWorkshop is conceived with the aim to unburden the meeting participants (the experts) by "us[ing] simple collaboration tools, thus minimizing potential technical problems and decreasing the time it would take to learn the tools" [32]. Besides simple collaboration tools, a support team oversees the meeting in a separate room to ensure that no problems occur and the meeting participants are not disturbed. To make the participants acquainted with the meeting topic and as a discussion guide, a pre-meeting information sheet is distributed and a pre-meeting training on the tools is offered. Conducting the eWorkshop [32] took about 2 hours and the chat contained 533 responses.

The eWorkshop is an application of *knowledge dust collector* in Basili's *Dust to Pearls* approach [33]. Dust represents knowledge spread immediately after it is captured (here in form of the meeting chat) and pearls represent the engineered experiences in form of ready-made guidelines. Figure 4.17 presents a general overview of the experience flow and documents.

Method: As main method, eWorkshop is a tool support for an anytime workshop. Secondly, eWorkshop supports collaboration by providing a tool to bring distributed participants together. The eWorkshop does not instantiate the experience as a by-product approach, though it roughly follows its principles. However, the fist goal to *capture experience during specific tasks within software projects* is not met. In the literature the workshop was specifically appointed to collect expert knowledge (about defect reduction methods) [32] and not within the usual software development tasks.

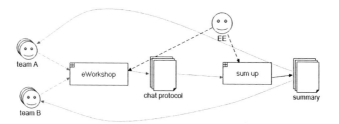

Figure 4.17.: Process, communication and documents with eWorkshop.

Light-weight criteria: C3 (keep sharing process short) and C5 (shift experience engineering away) are fulfilled. Through the usage of "simple collaboration tools" and the support team, the intention of the method is to decrease learning time and problems during the meeting for experience bearers [32] (C3). Much effort is shifted "'behind the scenes' regarding preparation, conducting, running, and analysis of the meeting" [33] (C5). The support team consists of five roles and has to follow a long list on how to set up and proceed in the meeting.

LIDs

Description: The Light-Weight Documentation of Experiences (LID) [212] is a workshop to capture project experiences. It is supposed to be organized "after a one-to-three month activity or project phase *of importance* is over" [212]. The idea behind LIDs is to collect "stories" from "key participants" during the meeting and protocol them in a document. The uttered statements should be written down as verbatim as possible, even including expletives. This way the emotional component of an experience is captured. LIDs does not require any preparation from the participants. They should only recount those events that they can recall at the time of the LID session. Only these experiences are worth collecting, as they must have been intense in a positive or negative way. The moderator uses a template with prepared questions. After the LID session, the document with the stories is put in a directory with other documents that were referred in the stories. Inside the experience document (the LID) links are created to the related documents at places they are referred to serving as an "access mechanism" [212].

An implementation of LIDs has been repeatedly conducted as part of the student software project course at the Software Engineering Group, Leibniz Universität Hannover. There, a LIDs session lasts about 1.5 to 2 hours. Since the course only lasts one semester (3.5 months), LIDs are conducted once after the project has ended. Figure 4.18 presents a general overview of the experience flow and documents.

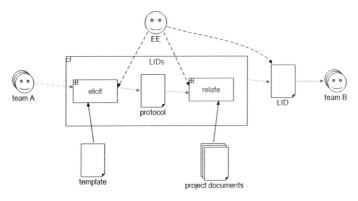

Figure 4.18.: Process, communication and documents with LIDs.

Method: Though practiced as a post-mortem workshop in the student course, in longer projects LIDs are supposed to be held every one-to-three months [212]. Thus it is defined as an anytime workshop.

Light-weight criteria: C3 (keep sharing process short) and C5 (shift experience engineering away) are fulfilled. C3 is covered, as "[p]eople usually like to talk about their recent adventures, so they will not consider it [the workshop] effort" [212]. Besides, participants do not have to prepare anything. Analysis and synthesis of the LIDs are conducted by the experience engineers (C5).

Light-weight Post-mortem Review (LPMR)

Description: A light-weight post-mortem review is a meeting "to collect information from the participants, make them discuss the way the project was carried out, and also to analyze causes for why things worked out well or did not work out" [87]. More precisely, the goal of the review is "capturing experience that might be useful for others". The workshop is guided by two researchers (or experience engineers), the one acting as secretary recording the workshop and the other overseeing the process. The workshop usually takes half a day and is conducted as a KJ-session [224], where participants write experiences on post-its, prioritize them and explain why this is a problem or success. Afterwards, the participants structure the post-its together in a root-cause diagram and discuss them on the whiteboard. After the workshop, the researchers write a report about the project and the issues revealed in the review. The report should be about

15 pages long. Figure 4.19 presents a general overview of the experience flow and documents.

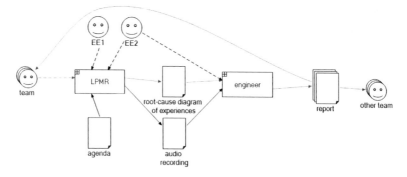

Figure 4.19.: Process, communication and documents with the Light-weight Post-mortem Review (LPMR).

Method: This is a post-mortem workshop, as it is an event where a group of people discuss their experiences and interact.

Light-weight criteria: C3 (keep sharing process short) and C5 (shift experience engineering away) are fulfilled. A major goal of the light-weight post-mortem review is to lower the time spent (C3): "it should not take much time for the project team to participate" [87]. Considering C5, an analysis of the review resulting in a report are done by the reseachers.

4.4.7. Experience Authoring

Answer Garden

Description: Answer Garden is a method to capture organizational knowledge in form of frequently asked questions and answers [10, 12]. Answer Garden approaches the problem that experts, e.g. in case of a hotline or generally in a company are often overburdened by a continuous stream of questions from, e.g. customers or new employees. The most questions are asked over and over again and only a few are new to the organizational knowledge base. Answer Garden unburdens experts by letting the knowledge seeker first answer "diagnostic questions" [10] in a tool that determine if the database already contains an answer to the initial question or if an expert can be contacted. Diagnostic questions aim at determining the exact problem like "Does the power light come on?" in the case of a PC malfunction. Besides answering diagnostic questions, knowledge seekers can browse the complete question and answer collection. Only if the question is still not answered, the knowledge seeker can personally contact the expert. Using the answers to the diagnostic questions, Answer Garden determines the appropriate expert for the question. In the implementation of Answer Garden new questions can be sent by email to the expert, though other sources are also possible [10]. The new question and answer are automatically saved in the knowledge base, if the expert answers the email. The answers can be kept as "simple text" directly from emails, but also in a more sophisticated presentation including structuring, styles, different fonts and graphics [10]. These modifications have to be done by the expert. Figure 4.20 provides a general overview of the experience flow and documents in Answer Garden.

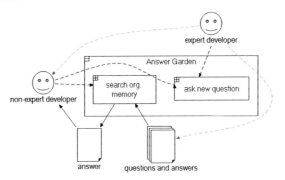

Figure 4.20.: Process, communication and documents with Answer Garden.

Though the implemented technologies and tools are not up to date, Answer Garden as a method to collect Frequently Asked Questions (FAQ) and unburden experts is still contemporary and can be directly applied to the software engineering domain. In the SE field, expertise location, identification and support is still a focus of research. Today, such approaches employ more contemporary methods like recommendation systems [174], support contemporary community experts like those in StackOverflow[4] [114] or examine social aspects [253]. Ye et al. also approach the problem that programmers are interrupted from their work by information seekers and thus become less productive [260]. They present a framework to "provide unified support for acquiring external information in code and documents with peers".

Method: Considering that the answers can contain experience and are written by experts, this approach is an instantiation of the experience authoring method. More specifically, each answer is can be regarded as a short experience report. Furthermore, the idea of Answer Garden is to

---

[4]http://stackoverflow.com/

create a collective memory in a distributed setting [12]. Thus it can be classified as collaboration support as well.

Light-weight criteria: C3 (keep sharing process short) is fulfilled. Experts are unburdened, as "Answer Garden eliminates the need to answer many simple questions [...] over and over again" [10].

Dandelion

Description: Dandelion is a Wiki plugin and is conceived for the purpose of collaborative authoring in a Wiki [58, 57]. Though the method instantiation does not explicitly mention knowledge or experience as an asset to collect, it is included in this thesis. Collaborative authoring is a general purpose, which can also mean creating co-authored experience reports in a distributed project. Besides, Dandelion was successfully utilized for collaborative report generation [57], which is close to an experience report.

Dandelion approaches two issues in collaborative authoring (of experiences) in Wikis. First, coordinative tasks like assignment of co-authors is not directly possible inside a Wiki. The coordinator must use other tools like email to contact co-authors, which means a context switch disrupting the work in the Wiki. Second, Wikis do not provide "effective" ways of activity and status monitoring around the authored texts [58]. Dandelion allows to add tags in the Wiki. The tags are executed by the plugin eliminating the need to switch to other. Dandelion then initiates a coordinative or awareness-enhancing task, e.g. emailing designated co-authors of an experience report in a distributed team. As an alternative to tagging inside the Wiki, Dandelion provides a "semi-structured, template-based approach that allows users to use templates to specify their requests in email" [57]. The method approaches the problem of a low Wiki acceptance being not part of the typical workflow in the group and cumbersome to access. Dandelion then processes the emails and generates Wiki pages and initiates coordinative actions. Figure 4.21 presents a general overview of the experience flow and documents.

Method: Dandelion falls in the main method category of experience authoring and secondary category of collaborative support. Though its main purpose is to support "collaborative authoring", this thesis considers the instantiation under the aspect of experience collection first.

Light-weight criteria: C1 (reuse known software), C2 (integrate into process and tools) and C4 (reduce cognitive load) are fulfilled. Dandelion reuses existing, well known and highly used email as means to share knowledge (C1). Dandelion is specially designed to support the existing collaborative authoring process around the Wiki and provides a way to add content to the Wiki though email, adapting to an existing communication process (C2). Dandelion fulfills C4. It is designed to minimize context switch interruptions for coordinative tasks integrating them in the Wiki and for opening it at the first place by allowing to submit content through email.

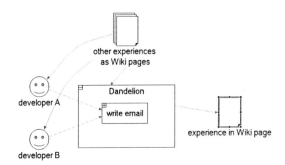

Figure 4.21.: Process, communication and documents with Dandelion.

Observation Sheet

Description: A quite simple way to externalize experiences at the time they happen is to write an observation sheet (also called *observation package* [230, 213, 148]). This is a one-page paper form (see [230], Figure 3 for an example of an observation sheet). An observation sheet usually consists of at least three input areas for the basic experience parts: observation, emotion and conclusion. When needed, more fields or checkpoints can be added [213]. It must be kept in mind though, that the main goal of this collection method is to keep the requirements and effort low and not to scare the experience bearer away with an overloaded form. Experience bearers are supposed to keep a stack of observation sheets in reach of his or her workplace to be able to quickly fill it out if something noteworthy occurs. An observation sheet usually contains very short experiences, just a couple of sentences for each experience part. An example of a real experience from a distributed development project [230]) observation sheet is displayed in example 4.1 [23].

Example 4.1 (Observation sheet).

Observation: *"Every time we are making a phone call, a lot is going on in the background. We are often disturbed during programming."*

Emotion: *"Annoyance: it is hard to concentrate."*

Conclusion: *"Could someone put a sign on the door and tell them to be quiet?"*

After some time (according to own experience after the project end, but can be more often during the project), the observation should be collected and engineered according to the process in section 2.1.3. Figure 4.22 provides a general overview of the experience flow and documents in an Observation Sheet.

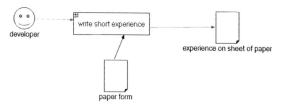

Figure 4.22.: Process, communication and documents with Observation Sheet.

Method: The method falls in the main category of experience authoring.

Light-weight criteria: C1–C5 are fulfilled. Observation sheets rely on paper and pen, which does not require learning (C1). Quickly writing down a couple of sentences on a sheet of paper is just a small change in the usual working process (C2). The main goal of the collection method is not to be "labor-intensive" [213] (C3 and C4). The experience engineer has no additional work after he has written down his experience (C5).

Mail2Wiki

Description: Mail2Wiki [113] is a method and tool support to facilitate writing texts in a corporate Wiki. The instantiation like *Dandelion* and *Wiquila* (introduced further down in this chapter) does not specifically mention experiences as asset to collect, but can be directly employed as it is for the purpose of experience writing into a shared Wiki. Mail2Wiki approaches problems related to Wikis: a "high interaction cost [. . .] for contributing and organizing content" and "poor integration with existing content-management tools and practices" [113]. Mail2Wiki provides three low-effort ways to contribute to a shared Wiki.

Mail2Wiki A The experience bearer can contribute a part of an email to Mail2Wiki, which is a Microsoft Outlook plugin. Outlook is considered "the most commonly used client among knowledge workers" [113]. Mail2Wiki provides a view, which suggests target Wiki pages as well as personal page favorites. Further, the plugin recommends a fitting section on the selected Wiki page. To facilitate contribution, the experience bearer can use familiar modes of manipulation like drag & drop text from an email directly into the target page. Mail2Wiki then appends the added content into the Wiki page. The experience bearer can afterwards edit the email and write additional text in the plugin view.

Mail2Wiki B With this mode of sharing, the experience bearer can contribute one or more emails without installation or configuration of a plugin. Mail2Wiki uses email as means to write Wiki content, which is the central content management and communication facility [89]. This method is similar to *Dandelion*. Thereby, the experience bearer has to add the Wiki page name as recipient, where the email content should be extracted to. The experience bearer can also send a batch of emails at once as attachment. There are other communication codes with the server (e.g. to get a Wiki target page recommendation) encoded in the recipient email address. The contributor then receives an email from Mail2Wiki with a Wiki page preview of his additions and edit his contribution.

Mail2Wiki C It is also possible to drag and drop a batch of email into the plugin.

Figure 4.23 presents a general overview of the experience flow and documents.

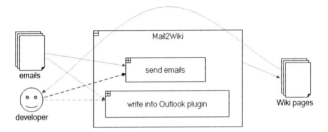

Figure 4.23.: Process, communication and documents with Mail2Wiki.

Method: The method category behind Mail2Wiki is experience authoring, as it directly supports adding new content by an experience bearer to an experience base.

Light-weight criteria: C1–C5 are fulfilled. Mail2Wiki reuses the common email system. It also reuses well known interaction strategies from the operating system like drag & drop for adding new content (C1). Especially Mail2Wiki B integrates well into the existing working process. It supports the usual process of communicating through emails (C2). All variants of Mail2Wiki save time and effort of the experience bearer (C3). Mail2Wiki B does not require any installation or configuration. The sharing process is very simple and quick just by writing a special email receiver. Mail2Wiki A requires more effort as the experience bearer has to decide which parts of his email he wants to share and explicitly drag the parts into sections of the target Wiki page. However, the decision where to paste the email text is facilitated by automatic suggestions. The author can quickly drag and drop the content into the chosen section. The immediate feedback saves time needed for corrections. Mail2Wiki minimizes the cognitive load by lowering the number of cognitive switches an experience bearer would have to make in order to contribute (C4). He does not have to open a new tool – the Wiki – but use his already opened email client. C5 is also fulfilled, as Mail2Wiki explicitly mentions a "curator" role which is separated from "contributors", the experience bearers. Curators have the task to "aggregate and package information on wikis", which is similar to experience engineering as defined in this thesis (see Section 2.1.3).

Wiquila

Description: Wiquila is a standalone Java client that supports authoring in Wikis and manage Wiki pages [202], relieving the author from using the Wiki itself. Similarly to *Dandelion* and *Mail2Wiki*, Wiquila approaches Wiki adoption issues in an enterprise environment due to lack of usability, low productivity due to syntax overhead, poor structure when content is growing, etc. (see Romberg [202] for the whole Wiki issue list or other research on Wiki problems, e.g. [43, 80, 128, 164, 179, 252]). The approach strives for features that ease and shorten the time for authors to write articles (e.g. experience reports). Wiquila provides "easy (assisted) insertion of links" to sources software project information inside and outside of the Wiki through context-sensitive suggestions. Wiquila allows content editing trough a WYSIWYG editor in a "word processor style" [202]. Many commands can be executed by shortcuts supporting power users.

Wiquila also strives for integration into Eclipse and Web browser, as well as desktop-like aspects like drag&drop actions. These aspects have not been implemented yet, though. Figure 4.24 presents a general overview of the experience flow and documents.

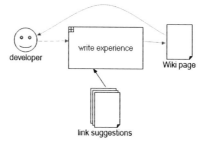

Figure 4.24.: Process, communication and documents with Wiquila.

Like *Dandelion* Wiquila does not explicitly mention knowledge or experience capture. Though Wiquila was conceived with the use case of authoring meeting minutes in mind, it is only "one of several use cases" [202]. It is possible to use Wiquila *as it is* for experience collection. This makes the difference between the excluded meeting minutes extraction tool (see Section 4.4.1) [179]. It was conceived *particularly* for the purpose of extracting tasks and decisions from meeting minutes. It cannot be used for other goals without a re-design of the tool.

Method: Wiquila is an experience authoring method, as it is its main purpose according to its description [202].

Light-weight criteria: C2 (integrate into process and tools) and C3 (keep sharing process short) are fulfilled. Wiquila is envisioned to be integrated into a desktop (drag&drop actions) and Eclipse (C2). To fulfill C3, the WYSIWYG editor allows a more easy way to enter text. Context-sensitive suggestions for links save time searching and typing. Shortcuts are intended to save time for power users [189, p.139].

4.5. Discussion

A threat to validity can be the probability to wrongly classify an experience collection method as light-weight. The concept of light-weight experience collection is not uniformly defined throughout the method instantiations. Many methods do not define and use a light-weight approach with the main goal to lower the effort for the experience bearer. For some, light-weight means *technically* easy to set up the tool support or as an *economic* solution. Both characteristics do not concern the experience bearer and do not unburden him in the experience sharing process. I did not include method instantiations that named their method light-weight in the above senses. An exception was made, if the technical setup process has to be done by the experience bearers themselves and is part of the knowledge collection process. This is the case for *eWorkshop* [32]. The method emphasizes "simple collaboration tools" [32]. Workshop participants (experience bearers) have to operate them during the knowledge-collecting workshop. Thus, this is part of the collection process and unburdens the participants.

The authors of the light-weight methods do not elaborate on the light-weightedness of their approaches in equal detailedness. Some authors put light-weight experience collection in the foreground (e.g. FOCUS, CRMT), while for others lowering effort is a rather marginal issue (e.g. e^3). Table 4.4 summarizes the light-weight methods and the focus of the publications.

- Some light-weight methods focus almost solely on specific algorithms, e.g. *Automatic Rationale Extraction* [199] or e^3 [51], and only marginally motivate that the approach should lower the effort for experience bearers. These cases are light-weight, though. Extracting knowledge from documents decreases the effort of explicit sharing through e.g. experience authoring methods, workshops or interviews.

- Authors of other light-weight methods describe a model or framework for a whole experience life-cycle (e.g. *Reflective Guides* [170]). Some other approaches cover rather abstract processes (e.g. most experience workshop instantiations) and do not emphasize specific tool support, as they are subsidiary or interchangeable. Here, elicitation of knowledge is only a small part and not emphasized.

- Other instantiations do not mention organizational aspects and mainly focus on specific tools (e.g. *Wiquila* [202], *Dandelion* [57]). I consider the methodology behind the tools in this thesis.

Often, but especially in the instantiations that mainly describe tools or algorithms, it stays untold who is responsible for which tasks in the experience collection and engineering process. It is often hard or impossible to determine if experience engineering is shifted away from experience bearers (C5 (shift experience engineering away) in Definition 4.3). It may differ depending on in which organizational framework the method is applied. In a setting similar to an Experience Factory [30] C5 would be fulfilled, in a self-organizing Community of Practice [251] it would be not. I consider C5 fulfilled, if it is clearly stated or the organizational setting of the method resembles an Experience Factory. Thus, if the authors state that their method employs experience engineers to administer or conduct the collection process, I assume that these experience engineers will also be responsible for further refinement of the collected experiences.

The Quality Circle (QC) method was also examined in the context of being light-weight according to Definition 4.3 and rejected as not light-weight. QC is an instantiation of an experience workshop method and originates from Japan in 1962. It became well known in the rest of the world in the 1980s [134, 232]. The Japanese version of a QC is a regular and voluntary meeting of employees, containing 5-12 members meeting once or twice a month [232]. QC members identify problems related to work and recommend solutions that management can approve or reject [134, 232]. A superordinate goal of a QC is to improve the team's spirit, increase their general productivity and lower grievances [134, p.126]. There are several reasons for not tagging QC a light-weight method. I have not found sufficient evidence in the literature (e.g. [134, 232, 263]) that QC has the goal to lower the effort and time spent for its participants. Dale and Hayward [73] and Hutchins [133] report that many QCs have failed (in the UK), among other things because the company underestimated the "effort and effect" that a QC demands and brings along [73]. A QC, though Hutchins claims their principle and methods as simple [134, p.43], nevertheless includes a rather long list of steps that knowledge bearers have to perform: brainstorming session(s), collect data and analyze it (with help of trained statistics), create cause and effect diagrams, verify solutions (the solutions should be a "thorough piece of work" [134, p.57]) and present results to the management. Also there is evidence that the QC method is a "scientific fad" [72, 225] of the 80s and is not relevant in contemporary research and industry

Table 4.4.: Overview of the light-weight methods (in the order they appear in Section 4.4) and the focus of their publications and descriptions.

Method Category	Instantiation	Environment	Algorithm	Tool(s)	Process
experience as a by-product	Collaborative Risk Management Tool [215]			+	++
	FOCUS [211, 215]			+	++
	Delta Analysis [214]				++
data mining	Email Expertise Extraction (e^3) [51]		++		
	Automatic Rationale Extraction [199]		++		
interview	Reflective Guides [170]	+			++
experience workshop	eWorkshop [32, 33]				++
	LIDs [212]				++
	Light-weight Post-mortem Review (LPMR) [87]				++
experience authoring	Answer Garden [10]			+	++
	Dandelion [58, 57]	o		++	+
	Observation Sheet [230]	o		+	++
	Mail2Wiki [113]	+		++	o
	Wiquila [202]			++	+

++ = main focus; + = less in focus; o = marginally mentioned; empty = not mentioned any more.

I also did not include *writing experience into a Wiki* as a method instantiation of experience authoring, though a Wiki is a well-known collaborative authoring tool [161, 34]. This also includes platforms like StackOverflow[5]. There are several reasons for this decision:

- I do not regard *writing experience into a Wiki* alone as a collection method, but just as a tool. It lacks a process what and how (much) to write. Light-weightedness, for example, depends on the length of the authored experience. Wikis in general are also very numerous in implementation, purpose, setup, style and plugins. These factors can also vary depending on the used Wiki version.

- According to literature, it is inconclusive, if writing experiences into a Wiki is a light-weight method according to Definition 4.3. Many literature sources report on adoption problems with Wikis (e.g. [252, 164, 80, 43, 128, 179, 113, 57, 202]). The acceptance of a Wiki often depends on the culture in the organization. I observed this problem participating in the collaborative project *e performance* [179]. On those grounds I do not consider *writing into a Wiki* in this overview. This is a debatable decision. Conducting a quantitative assessment of the here presented light-weight methods (in Chapter 5), I list this tool support as a heavy-weight method, including assumptions about the experience content and size. The results of the measurement show that this method is borderline light-weight (see Figure 5.5 and Section B.2.2 for a description of the method), though.

- *Wiquila* is a method that is supported by a Wiki with a plugin that facilitates authoring in the Wiki. This method reflects one possible specialization of the more general activity of *writing experience into a Wiki*.

Moreover, the catalogue of found light-weight experience collection methods may not be complete. Some method instantiations may have been overlooked or considered as not relevant because they either were not explicitly mentioned as knowledge or experience collection methods or were not sufficiently described in the publication. Finally, the search methodology is not as systematic as prescribed by Kitchenham et al. [151] lacking a protocol of retrieved literature and the databases.

4.6. Contributions

This chapter presents the first part of the light-weight experience collection framework answering the research question **RQ2:** *How can light-weight experience collection methods be identi-*

---

[5]http://stackoverflow.com/

fied?

It provides five criteria of light-weight experience collection derived from the definition and a literature review (rephrased): C1) reuse or extend existing software; C2) integrate into existing tools and processes; C3) keep time to share short; C4) minimize experience bearer's cognitive load and C5) shift experience engineering activities away from the experience bearer. These criteria can be used for assessment of light-weightedness, but also as a guideline to create a light-weight experience collection method.

Furthermore, this chapter provides a catalogue of existing light-weight experience collection method categories and method instantiations for each category. These method instantiations are analyzed according to the defined criteria providing rationale for their light-weightedness. In the next chapter, this catalogue and the derived criteria are used as basis to determine a light-weightedness portfolio based on a constructed measurement system.

5. Measurement System for Light-weight Experience Collection

The last chapter provided a qualitative assessment of light-weightedness of experience collection with a catalogue of suitable methods. This chapter defines a measurement system to quantify it.

In order to answer research question **RQ3:** *How can light-weight experience collection methods be assessed?*, this chapter presents indicators enabling an experience engineer or manager in a global software engineering endeavor to

- provide metrics for measurement of an experience collection method and

- analyze the light-weight collection methods from Section 4.4 according to the measurements, contrast them to heavy-weight methods and provide a portfolio and gradations of light-weightedness.

5.1. Measurement Perspectives

The two most important factors (see sharing barriers in Chapter 3) that have to be compared when measuring or assessing an experience management (EM) initiative are the *effort* and *benefit* [212]. This thesis differentiates four perspectives when measuring effort and benefit [216, p.41] (from small to large):

Personal: Benefit and effort factors that concern the experience bearer personally.

Team: Benefit and effort factors that concern a team within a collaborative project.

Organizational: Benefit and effort factors that concern the partner organization a team belongs to.

Inter-organizational: Benefit and effort factors that concern all partner organizations within a collaborative project.

Since light-weight collection methods focus on supporting the experience bearer on personal level, effort and benefit must be considered on this level. On personal level, there are mainly two types of experience bearers: a *learner* and a *provider*. The definition of the learner role (Definition 5.1) is based on Schneider [215]. A learner defines a person who may *potentially*, but very likely seek new knowledge. Thus, a learner is a potential user and beneficiary of the knowledge base. According to the definition, a learner must not necessarily be an experience bearer (with relevant experience for the EM initiative), but he may become one at any time during the project.

Definition 5.1 (Learner).
A learner is a person who will need to acquire new knowledge and experience.

The definition of a provider (Definition 5.2) is based on the experience base roles and views [23]. Both roles aim at distinguishing those project participants, who provide experience (provider) and who may need it (learner). As stated in previous chapters, an EM initiative and especially the collection method must also be beneficial for the provider.

Definition 5.2 (Provider).
A provider is an experience bearer who has already contributed to the EM initiative by sharing his experience.

5.2. Derivation of Measurement Criteria

Effort is defined as

Definition 5.3 (Effort).
Effort is a "specific and quantifiable count and/or measure of definable labor units that it is deemed [...] to be required in the attempts to arrive at completion of a phase (or of the entirety) of a particular schedule activity and/or work breakdown structure component, a distinct control account, or the project as a whole" [6].

There have been many attempts to define and measure work or task effort (e.g. [249, 208, 139, 61, 99]). A way to quantify effort is to measure the time on task(s) [91, 233]. In the context of determining the light-weightedness of an experience collection method, duration is not sufficient. Definition 4.3 suggests that there are several ways to lower the effort. Based on this definition, four factors can be identified that influence the overall effort and two factors for benefit. Table 5.1 summarizes the influencing factors to the overall effort and benefit. In the following, each factor is shortly described. In the next sections they are presented in detail.

Table 5.1.: Factors influencing the effort and benefit of an EM initiative. Each factor can be estimated through indicators and is relevant for the displayed perspective.

Main factor	Influencing factors	Perspective	Indicators/ Metrics	Sec.
effort				5.3
	↓ experience engineering (EE)	org., inter-org.	duration of EE process, number of EE tasks to do, experience rawness, experience maturity	5.4.1
	↓ administration	org., inter-org.	number and duration of administrative tasks	5.4.2
	↓ time to share	personal	sum of atomic step duration, frequency, participants, shared expriences	5.3.1
	↓ cognitive load	personal	concepts to recall, mental demand, tool and process integration, ad hoc sharing	5.3.4
benefit				5.5
	↑ immediate perceived benefit (without EE)	personal, team	experience rawness, experience maturity, saved time with EM support	5.5.1
	↑ possible benefit with EE	all	improved process	5.5.2

Experience engineering (EE) effort: For an experience collection method to be light-weight, experience engineering should be shifted away from the experience bearers (C5 in Definition 4.3). This means that someone else has to do this task. For the overall effort assessment this is also an influencing factor.

Administration effort: This effort is not explicitly mentioned in Definition 4.3 as one of light-weightedness criteria to minimize. Apart from the task of engineering the captured experiences, the tools or processes employed during the EM initiative have to be administered and maintained. This affects organizations in a collaboration.

Time to share: According to C3 from Definition 4.3, the sharing process should be short. This can be measured by the time the experience bearer invested for sharing his experience.

Cognitive load: Criterion C4 demands that the cognitive load on the experience bearer should be minimized. A low cognitive load would facilitate the experience sharing activity.

Considering the benefit and according to the sharing barriers (Figure 3.1), an EM initiative must provide useful content and immediate benefit to be successful. Thus, it is reasonable to provide means to measure benefit. The overall benefit of an EM initiative is divided into two components:

Immediate perceived benefit (without EE): This is the benefit perceived by project participants (learner and provider from the collected experiences without engineering it.

Possible benefit with EE: This is the benefit the collected experiences return after they have been engineered. This type of benefit can be long- or short-term.

The following sections provide a detailed description how to measure effort and benefit. Section 5.3 presents measurement of personal effort, Section 5.4 presents organization effort or cost, Section 5.5 introduces benefit metrics and Section 5.6 combines the relevant metrics to present the overall metric for light-weightedness of a collection method.

5.3. Measurement of Personal Effort

This section presents scales to measure factors influencing *effort*.

5.3.1. Time to Share

Time needed to participate in one experience collection session is the sum of atomic steps during the sharing process. An *atomic step* is an action, e.g. a click on a link, window switch, login action to a site or write a text. The overall invested time t_M to share *one* experience for *one* experience bearer with method M within a predefined *time period* is expressed as

$$t_M = \begin{cases} f \cdot p \cdot \sum_i t_i, & \text{if } e = 0 \\ \frac{f \cdot p}{e} \cdot \sum_i t_i, & \text{else} \end{cases} \tag{5.1}$$

with

$t_i \in \mathbb{R}^{\geq 0}$: time of an atomic step or task i that is part of one session of experiences sharing with the tool support or method,

$e \in \mathbb{N}^0$: number of experiences collected during a given time period (e.g. project duration),

$p \in \mathbb{N}^0$: number of experience bearers (except experience engineers) that conducted or were part of the collection session and were the sources of collected experiences e,

$f \in \mathbb{N}^0$: number of executions of the collection session during a given time period.

The duration of an atomic step can be characterized by either the actual time the knowledge bearer needs to interact with the system or by an imposed time constraint. A constraint can be, e.g., a pre-defined workshop duration. It is assumed that each participant contributes equally to share e experiences and needs an equal amount of time for similar sharing activities.

The amount of elicited experiences shared during one session can vary. In an experience workshop, one participant will likely contribute more than one experience, in contrast to an observation sheet, which is designed to capture one short piece of experience. In order to calculate and compare the time to share *one* experience for one person, the time must be normalized. The time to share is divided by the number of (potential) raw experiences e captured during the pre-defined time period. In many cases, e is a collective result of more than one experience bearer. Dividing the time to share by e would mean that only one person contributed e experiences. Thus, the time to share must be multiplied with the number of authors (p) of e. The default team size is assumed 5 (see Table 5.2). Finally, the different collection methods not only vary in the time to share of one session but also in the frequency they are carried out.

Table 5.2.: Prerequisites and assumptions for time to share calculation.

Default prerequisite	Value	Rationale
team size	5	average team size [185]
project duration	2 years	average project duration [185]
working state at the workplace	PC started, open IDE	assumption of a usual software development use case

Take an example of where an experience authoring method A may require only minutes for writing a short experience and a method B as a half day workshop. However, method A is expected (based on e.g. evaluations or experience) to be executed several times a week, while method B only twice during a project. Neglecting the frequency in the overall calculation would make a comparison of the collection methods with each other impossible. This mismatch is relativized by multiplying the frequency f of (intended) execution of each method during a defined time period (here, 2 years). It is assumed that each experience bearer shares experiences constantly during the defined time period.

Figure 5.1 visualizes the described relations and the need for normalization of the time to share. The upper part of the graphic displays the times to share of two exemplary collection methods A and B and their steps. The methods differ in overall time (width of the rectangles), number and duration of their steps, number of participating experience bearers and the outcome experiences. The bottom part of the graphic (beneath the gray dashed line) schematically visualizes different frequencies and durations of both methods. Without normalization by the output of both methods, i.e. the number of overall experiences produced, it would not be possible to compare their time to share. Example 5.1 presents a sample calculation of the sharing time.

Some collection methods may pose an exception to Formula 5.1. They plan for several subsequent sharing sessions that differentiate in process, duration and outcome. In this case, t_M is the sum of sub-sessions' times to share:

$$t_M = \sum t_{Mj}, \qquad (5.2)$$

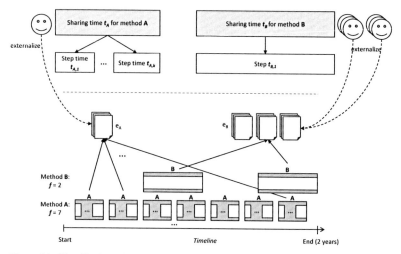

Figure 5.1.: Visualization of the components that influence the time to share of one experience.

where $j \in \mathbb{N}$ is the number of sub-sessions of M.

Example 5.1 (Time to share).

Assume an experience collection method M, where one team of 5 software engineers takes part in an experience workshop that lasts 3 hours. This experience workshop is conducted 4 times during a 2-years project. The outcome of the workshop is an experience report containing 30 experiences. The report is written by the experience engineer. The overall amount of collected experiences during the project would be $e = 30 \cdot 4 = 120$. This means that $p = 5$, $t_{M,1} = 3h$, $e = 120$ and $f = 4$. This collection method contains 2 atomic steps: 1) participating in a workshop and 2) writing the report. As event 2) is not conducted by experience bearers, it is dismissed in this equation. The time to share for one participant using the method M would be $t_{M,1} = \frac{4 \cdot 5}{120} \cdot 3h = 0.5h$.

An example would be *Reflective Guides* (see Section 4.4.5). The sharing process with this method consists of two independent sub-session that differentiate in their duration and outcome: 1) filling out the guides, i.e. questionnaires and 2) participating in a workshop. For the first activity, there is only one participant, while the second sharing activity is a team meeting. The filled out questionnaires from sub-session 1 are then discussed in sub-session 2 – the experience workshop. This may result in some new insights or trigger other experiences from the participants.

Therefore, both activities must be evaluated separately and summed up to receive the overall time to share. A detailed calculation of the time to share for *Reflective Guides* is presented in Appendix C. The rationale for the assumed values in the calculation are presented in Appendix C.1.

Step Duration

In some cases it was not possible to obtain information on the duration of each step needed to perform a collection method. In this case, to provide a basis for estimation, I define default durations based on literature and experience. Table 5.3 displays common atomic steps or tasks that can be part of an experience collection process with their estimated average duration. The rationale behind the durations is described in the following.

Table 5.3.: Common atomic steps and estimated time to accomplish them.

Activity	Approx. duration
click on a button, link or menu	2 sec
open a new program	6 sec
write a short observation sheet (ca. 5 sentences)	5 min
write one page (ca. $1,800$ characters) of free text	45 min
read one page (ca. $1,800$ characters)	3 min
conduct a focused (semi-structured) interview	30 min
conduct an experience workshop	1.5 h

Although many values are not empirically evaluated, if applied to all examined methods equally, the time to share of the methods is comparable. The estimations are based on the following assumptions and constraints:

- The written language is assumed German. It is the language the case studies participants spoke and wrote (see Section 6). Other languages have different characteristics like sentence length[1], which would yield other values.

- The language of texts to read is assumed to be familiar to the reader, i.e. have a familiar complexity and be from his field of work. It should not slow him down notably.

- I further assume that the experience bearer – a software engineer – is very well acquainted with the operating system and PCs in general. He is not physically disabled.

---

[1]English language has an average sentence length of 15-20 words [160, 2], while German has 7 [38].

- I assume a commonly used graphical operating system (OS) with mouse and keyboard. Though the concepts in this thesis are OS-independent, some assumptions and calculations, like the default durations, may differ depending on the used OS (and OS version). Also specific user interaction processes, e.g. selecting a document menu, can vary. In cases, where user interaction and resulting effort depend on the used OS, I assume MS Windows 7.

In the following, I elaborate on the rationale behind the assumptions of the chosen values in Table 5.3.

Click on a button, link or menu: The duration of a simple interaction like a click on a link or button is based on an evaluation of user selection methods (click and hover) [40]. I use the time to click on a single link in this case study and round it up.

Open a new program: Opening a new application, should take about two or three times as much time as clicking on a link. I also assume that there are two common ways to open a new application. First, I consider the case where the experience bearer has a program shortcut on the desktop and has another window open and in focus (see assumed default setting use case of a developer in Table 5.2). Then, he needs to 1) minimize the current window (a click, 2 sec) and 2) search for the program on the desktop (assume about 4 sec.). In the second scenario, the person would search for the program in the program menu, which takes approximately the same time. This estimation does not consider any delays at program start.

Write a short observation sheet : The time for externalizing a short piece of text is the duration to write a short observation sheet. This value is based on observations during the evaluation of the Observation Widget with Heuristic Support [22, 24] in Section 6.2. On average, participants needed 10min to write 2 experiences consisting of about 5 sentences each including thinking. This takes 5min to write one experience of 35 words, considering that an average sentence of prose has approximately 7 words [38].

Write one page (ca. 1,800 characters) of free text: The time to write a longer piece of text (one page) is an extrapolation of the above measure. A standard page (for writers) contains $1,800$ characters for German language [156]. The average word length is 5.1 characters in German [4]. This makes a rough estimate of $1,800/5.1 \approx 305$ words on a standard page. Compared to a short observation sheet, a standard page contains $305/35 \approx 9$ times more words. Thus, writing a standard page takes approximately $5\text{min} \cdot 9 = 45\text{min}$, assuming a constant and equal writing speed.

80

Read one page (ca. 1,800 characters): For the time to read, I presume a reading speed of average readers as basis, which is about 200 words per minute with $> 80\%$ rate of understanding [158]. A page with 305 words would take 1.5min to read. Assuming that software engineering documents need more careful reflection and 100% understanding rate, I round the value up to 3min

Conduct a focused (semi-structured) interview: An indicator to a focused interview duration is given by DiCicco-Bloom and Crabtree [85].

Conduct an experience workshop: The default meeting or experience workshop duration is based on a typical meeting in an American corporation [201].

5.3.2. Experience Maturity

When raw experiences are collected, they can have different levels of maturity depending on the collection method and effort made during collection[2]. Experience maturity influences the engineering effort and immediate benefit for providers and learners. The higher the maturity of collected raw experiences is, i.e. the more pre-engineered they are, the less effort the experience engineer potentially needs to create highly reusable experience artifacts. The main goal of experience engineering is to raise the maturity of collected experiences and thus to make them reusable.

The notion of experience maturity has been defined in the literature, e.g. by Maier et al. [166] and Attwell et al. [20]. Although their definition is applied to knowledge in general, it can also be applied to experiences.

This thesis uses an adapted version of Maier et al.'s notion of maturity and consider experience *legitimation* as an indicator. Legitimation (or commitment) is the level of support the experience has received. This commitment can stem from groups, teams, experience engineers or communities. It has influence on the reliability of an experience and indirectly reflects the passed stages of experience engineering (see Figure 2.4 in Section 2.1.3). Maier et al. additionally define three further factors that influence maturity [166]:

- *Hardness*: The (alleged) validity and reliability of experiences.

- *Interconnectedness*: The degree of connections to other topics (*not* the degree of linkages to the original context of the experience).

- *Teachability*: More mature experiences are more easy to teach to others, while less mature experiences are less teachable.

---

[2] At this point, it is irrelevant who makes the effort.

Table 5.4 presents the maturity scale based on legitimation. A higher legitimation value means higher maturity. The other three above mentioned factors correlate with legitimation. Besides, they are directly or indirectly reflected in the experience rawness, which is introduced in the next section.

Statements for maturity 4 and 5 are based on previous work, where an experience engineering process for globally distributed projects is presented (see Section 6.3). There, engineered recommendations (maturity 5) are shared among all partner teams, while the more sensitive experiences (maturity < 5) can only be viewed and commented by the team that authored them.

Table 5.4.: Experience legitimation as indicator for experience maturity (adapted from [166]).

Maturity	Legitimation
1	not available
2	experience bearer, possibly confirmed by few colleagues
3	colleagues and possibly experience engineer(s)
4	experience engineer(s) and project team
5	experience engineer(s) and (all) teams from collab. partners
6	experience engineer(s) and specialized experts (e.g. course vendor, process owner)

5.3.3. Experience Rawness

Besides experience maturity, the experience rawness factor determines how much the collected "raw" experience is already pre-engineered, i.e. experience bearers have conducted experience engineering tasks as presented in Section 2.1.3. Experience maturity has an influence on the rawness factor. While tasks $1 - 3$, 5 and $6 - 8$ focus on an appropriate presentation, writing style and structure of the collected experiences, tasks 4, 7 and 9 directly concern the experience maturity. In task 4, the experience engineer looks for recurring experiences, thus trying to validate if the experience has been confirmed. In task 9 the experience engineer (or a different expert) has to maintain the recommendations and promote them to "best practices", if they have been confirmed by practice as such. Steps 4 and 9 raise the experience maturity. Thus, an implication that *low rawness factor usually induces a high maturity factor* can be assumed. But the other way round, it does not have to be the case. An example would be the experience workshop method *LIDs* [212]. It results in a protocol that has a rather high rawness being a direct recording of what was said. The maturity of the externalized experiences is usually medium or high because the experiences are discussed in the group, which is also recorded in the minutes document.

Table 5.5 provides a scale to estimate rawness of collected experience artifacts. A high raw-

Table 5.5.: Experience rawness scale with experience characteristics and experience engineering (EE) tasks according to Section 2.1.3. ↑ marks characteristics of rawness r that have positively changed towards a lower rawness compared to $r + 1$.

Rawness score	Characteristics	EE tasks from Sec. 2.1.3
6	*form:* experience is implicit and not directly distinguishable from other content *style:* badly readable, incomplete, highly contextualized, suboptimal granularity and size, can contain sensible information *maturity:* 1 – 2 *examples: annotations, emails*	1 – 9
5	*form:* ↑ contains narration of observed events or recommendations *style:* badly readable, incomplete, highly contextualized, can contain sensible information, suboptimal granularity and size *maturity:* 1 – 2 *examples: paper observation sheet, informal feedback sheet, informal complaint*	1, 3 – 9
4	*form:* ↑ narration contains experience as triple (observation, emotion, conclusion) *style:* ↑ rephrased, ↑ complete and traceable, highly contextualized, can contain sensible information, ↑ better granularity and size *maturity:* 1 – 2 *examples: complete, readable, traceable observation sheet*	1, 4 – 9
3	*form:* ↑ procedures to follow as full-text *style:* rephrased, complete and traceable, ↑ reusable and context clear, suboptimal granularity and size, ↑ categorized *maturity:* 3 – 6 *examples: lessons learned document, experience report*	1, 4, 6, 9
2	*form:* ↑ procedures to follow *style:* ↑ rephrased, ↑ complete and traceable, highly contextualized, can contain sensible information, ↑ harmonized *maturity:* 3 – 6 *examples: checklist*	1, 5, 9
1	*form:* ↑ procedures to follow (process) *style:* categorized, complete and traceable, reusable and context clear, ↑ anonymized, ↑ harmonized *maturity:* 5 – 6 *examples: company rules, compulsory guideline*	9

ness means that more experience engineering effort is needed. A low rawness score means that most of the experience engineer's work have already been done. The rawness score reflects the amount of experience engineering tasks that have to be performed. It also indicates the extent of benefit the artifacts would provide to readers. Reading experience artifacts of high rawness, e.g. a stack of annotations or a batch of emails (rawness 1), would be very time consuming. Artifacts of very high rawness contain a lot of information that is not experience. An example would be annotations mostly denoting grammar corrections or long, irrelevant emails about various topics. A method to collect and share project document annotations is presented in Section 6.1.

Summing up, Figure 5.2 schematically illustrates the notions of experience *maturity* and *rawness*. It reveals that experience of low maturity is anecdotal. Gaining proof and support it becomes more mature. Experience of a high rawness score is raw. After engineering, it becomes more elaborate and refined.

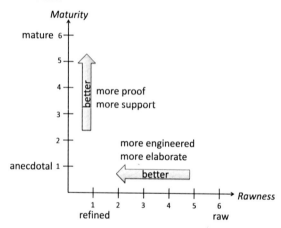

Figure 5.2.: Experience maturity and rawness.

To estimate the granularity and size that is included in the rawness calculation, the next section provides metrics for both notions.

Granularity and Size

Experiences can be of different granularity and size. Both notions are interconnected, but can be defined separately. This section defines both terms, illustrates the interconnection between them and provides a scale to estimate the granularity and size of experiences.

In Granularity Computing research as part of the Artificial Intelligence research field, *granularity* means the "measure of refinement degree of knowledge and information" and is the degree of information abstraction [123, 264, 165, 144]. Definition 5.4 builds on this granularity concept in the Granularity Computing research area.

Definition 5.4 (Experience granularity).

The experience granularity is the grade of an experience artifact's detailedness and thus its distinguishability from other experience artifacts.

A fine-grained experience artifact is detailed, has more context and is very likely distinguishable from other experiences of the same topic, while a coarse-grained one is vague, leaving details and context out, abstracting them away [144, 264]. Table 5.6 presents a granularity scale, which is used in this thesis for estimation.

Table 5.6.: Experience granularity scale.

Granularity	Metric
coarse	very abstract description; broad or no context about project, environment, tooling and people
mixed	either contains vague and very specific descriptions, or description of medium vagueness
fine	very specific description; narrow context about project, environment, tooling and people

Furthermore, Examples 5.2 and 5.3 illustrate the notion of granularity.

Example 5.2 (Coarse-grained observation).

Team member: *"We had a problem with an agile process."*

Project leader: *"An agile process did not work well for us."*

Example 5.3 (Finer-grained observation).

Team member: *"We had a problem with the agile process. We were forced to do Scrum but without a Scrum Master. We were often imposed new requirements during a sprint!"*

Project leader: *"An agile process did not work well for us. The team still had to write a specification for the contract beside user stories. This was extra work that delayed the project."*

Example 5.2 contains two pieces of experience that have a rather coarse granularity. They are vague and do not convey specific details or explanations. Thus, their content is indistinguishable from each other: it is only clear that both software engineers criticize an agile process. On the other hand, Example 5.3 presents a finer grained version of these experiences. They convey more information stating different problems in agile development and thus are distinguishable from each other.

Experience size is defined as follows:

Definition 5.5 (Experience size).

The experience size of an experience artifact is the number of atomic experiences (observed events). This number influences the length of the experience artifact.

This thesis uses the size scale in Table 5.7.

Table 5.7.: Experience size scale.

Size	Metric
very long	more than 5 pages
long	1 to 5 pages
medium	$1/2$ to 1 page
short	a couple of sentences or less

I distinguish nine different cases of relations between experience granularity and size, which are depicted in Fig. 5.3. An experience artifact, outlined by a black circle, can be of coarse (the cloud) or fine (the small circle) granularity and report of one or many observed events (number of clouds or circles).

An experience artifact can consist solely of coarse-grained statements and be of big, medium or of small size (see Table 5.7). Both experiences from Example 5.2 fall into category (3). An *enumeration* of vague statements like Example 5.2 could be case (1). There can also be a long, medium or short fine-grained experience. A long, detailed experience report is case (4). A statement like the one in Example 5.3 is case (6). An company-specific acronym could also be an example for (6) [216, p. 62][3]. Often, an experience artifact is of mixed granularity (cases (7)-(9)). This can be a long (7), medium (8) or short (9) abstract recommendation having a concrete example.

Considering the same atomic event, it can be assumed that the size of a fine-grained experience is usually bigger than the same experience with a coarse granularity. The first grants more details than the second. The first experience from Example 5.3 has 31 words. The more vague version of the same experience (Example 5.2) consists only of 8 words.

---

[3]Schneider uses granularity as a notion of experience artifact size (see Definition 5.5). There, he does not explicitly mention experience abstractness as defined in Definition 5.4, though.

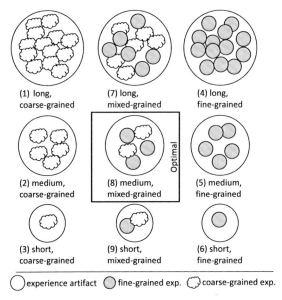

(1) long, coarse-grained	(7) long, mixed-grained	(4) long, fine-grained
(2) medium, coarse-grained	(8) medium, mixed-grained	(5) medium, fine-grained
(3) short, coarse-grained	(9) short, mixed-grained	(6) short, fine-grained

◯ experience artifact 🔘 fine-grained exp. ☁ coarse-grained exp.

Figure 5.3.: A model displaying interconnections between experience granularity (fine, coarse and mixed) and size (short, medium, long). The black circle encapsulating the granules represents an experience artifact as a whole. The blue cloud granule represents a coarse-grained (vague), while the yellow circle is a fine-grained (concrete) description of an experienced event.

Hypothesis 5.1 (Optimal experience size and granularity).

An optimal experience should be of medium size and mixed granularity.

An optimal experience according to Hypothesis 5.1 is case (8) in Figure 5.3 [216, pp.62, 140]. Long experience narrations can be too tedious and time consuming too read. This could pose a high mental demand and overburden the reader (see Section 5.3.4 for cognitive load and its influencing factors). Thus, an experience should be as short as possible. However, a very short experience is often not substantial enough [216, p.62]. An experience should contain enough context to understand it but be abstract enough to transfer and reapply it to other situations. Very fine-grained experiences may be too specific to reapply, while very coarse-grained ones could be too vague [216, p. 140]. An optimal experience should contain both types of experience granularity. This would be a mixed-grained granularity of medium length.

Assessing experience granularity and size of collected experience is often difficult and almost never exact. This thesis estimates the ability, opportunity and motivation of the experience bearer to share experiences of a certain granularity and length. Almost all estimations are assumptions and tendencies towards or likelihood of certain granularities and sizes. Sometimes it is only possible to exclude certain granularities or sizes.

5.3.4. Cognitive Load

While Section 2.2 provides an introduction on cognition and learning, this section operationalizes cognitive load, deriving metrics to quantify it.

Cognitive load influences how demanding a task can be for a person [240]. In the context of this thesis, the task is *experience sharing* by a software engineer. A high cognitive load means that the sharing process would be difficult and thus requires more effort. Therefore, the goal of a light-weight collection is the reduction of the cognitive load.

Cognitive load has been subject to research in different research fields. Two main fields are Human Computer Interaction (HCI) and Computer Supported Cooperative Work (CSCW) (e.g. [110]) in the context of improving usability (e.g. [189, 68]) and support (multimedia) learning and work [240, 171].

Beside usability engineering and learning, cognitive load can be directly applied to experience sharing. It is influenced by various factors that can be summarized in three main factors with more concrete and measurable sub-factors:

1. Task complexity [171, 240]

    a) Concepts or rules (*not* experiences) to recall

    b) Mental demand of task(s)

       - Amount of reading or writing

       - Amount of analytical thinking

2. Support of the user's mental model and his expectations of system behavior through prior experience [189, 110, 68]

    a) Integration into present tooling environment

    b) Integration into or change of work processes

3. Ability of ad hoc sharing [110]

Task complexity depends on the amount of concepts or rules the experience bearer has to recall. Concepts or rules can be a special language syntax or process steps, e.g. to recall a formula or the steps in the Goal-Question-Metric (GQM) method [244]. Task complexity does not consider the recollection of the experience itself, but other information that is needed to perform the sharing. Further, task complexity depends on the mental demand of the task. Mental demand includes the amount of reading and writing the experience bearer has to perform as well as the amount of analytical thinking required to perform the sharing task. A task that requires analytical thinking could be, e.g. to fill out a GQM abstraction sheet or create a cause effect diagram. The second main factor is the support of the experience bearer's mental model (see e.g. Cooper et al. [68, pp.27-41] on an introduction on mental models) and his expectations of the system's behavior. This can be supported by integrating the task and the tooling needed into the experience bearer's present working environment and his processes. The third factor is the opportunity to externalize experiences at the time they happen. According to Grudin, a lot of human problem solving is ad hoc and not post hoc[4] [110]. Thus ad hoc procedures and arrangements should be encouraged. This can be transferred to experience sharing. If ad hoc sharing is impaired or not possible, the experience bearer would have to remember the experience until the moment he is able to share it. This would mean a cognitive strain. Some methods only partly allow ad hoc sharing through reflection-in-action, e.g. during an experience workshop.

Beside the above mentioned, there are psychological factors involved like motivation or laziness, anticipation, skill and fatigue [177, p.5]. These factors can complicate the cognitive load estimation greatly. Meshkati and Hancock refer to two research papers discussing these problems [177, p.5]. In this thesis, I assume that the average software engineer is not lazy, but also not an altruist [216, pp.157-159] and is not unduly fatigued.

The measurement scales for characteristics 1a), 1b), 2a), 2b) and 3 are displayed in Tables 5.8–5.12. Since the estimations are rough, I chose a three-level scale for simplicity. In case of multiple indicators for a score, these indicators should be considered as connected by a Boolean OR. Table 5.13 presents the scale for the overall cognitive load, which is the sum of the values from Tables 5.8–5.12.

Table 5.8.: Concepts or rules to recall (1a).

Score	Indicator
2	Unfamiliar language syntax or a long list of rules to recall
1	Very few rules or terms to recall
0	No unfamiliar language or terms to recall

In case where the experience collection method plans for several successive activities that

---

[4]Post hoc would be, for example, a post-mortem workshop months after the experience happened.

Table 5.9.: Mental demand of a task (1b). See Table 5.7 for experience size scale.

Score	Indicator
2	1) Read or write a long text 2) Read or write a medium, but sophisticated text 3) Requires much (analytical) thinking
1	1) Read or write a short text 2) Requires some (analytical) thinking
0	1) No reading or writing 2) No (analytical) thinking required

Table 5.10.: Integration into present tooling (2a).

Score	Indicator
0	1) A plugin of an existing (and well known) tool 2) Sharing with or through an existing tool 3) No tool required
1	New standalone tool, but uses common practices and behavior
2	1) A standalone tool with unfamiliar behavior and interface 2) Method requires special tooling

Table 5.11.: Integration of the sharing process into work process with an experience collection tool or method (2b).

Score	Indicator
0	Does not change or disrupt work process
1	Usual work process (step) must be insignificantly altered
2	Added or new activity or event

Table 5.12.: Opportunity of ad hoc sharing (3).

Score	Indicator
0	Can immediately share experience at the time it occurs
1	Can share ad hoc experience only during a certain time span
2	Cannot share experience in an ad hoc manner

involve a different number of participants and produce different experiences (e.g. *Reflective Guides*), cognitive load should be the maximum of the cognitive load scores for each activity (see equation 5.3). This way the "bottleneck" of the collection method is taken into account. The cognitive load c_M of an experience collection method M is the maximum of the cognitive

Table 5.13.: The overall cognitive load metric.

Score range	Cognitive load
9 - 10	very high
7 - 8	high
5 - 6	medium
3 - 4	low
1 - 2	very low
0	none

load values $c_{M,i}$ of its non-atomic activities:

$$c_M = max(c_{M,i}), i \in 1, \dots, n \tag{5.3}$$

5.4. Measurement of Organizational Effort

An important management barrier to a successful implementation of an experience management (EM) initiative, is the cost of the initiative (barrier 30 in Table A.1). It is not included in the cause effect diagram in Figure 3.1 because it is an organizational barrier and not an issue on personal level of an experience bearer. However, the cost of the EM initiative is relevant if a knowledge manager has to decide if an EM initiative is cost-effective. This barrier can directly influence personal sharing behavior. An overly expensive initiative would not be fully supported with all the features and would be cut down expenses. This may either raise the effort (e.g. by imposing EE activities on experience bearers) to share or diminish the benefit. This section investigates the costs that are directly related to the collection methods. Organizational costs are reflected through effort and man hours an experience engineer spends on *experience engineering* and *administrative tasks* for the EM initiative.

5.4.1. Experience Engineering Effort

This section defines a scale of experience engineering effort according to the process described in Section 2.1.3. Depending on the type of the collected experiences, the effort to engineer them can vary.

The experience engineering (EE) effort can be expressed as time needed to perform the tasks from Section 2.1.3. The time function depends on rawness and maturity of collected experience. However, it stronger depends on the rawness than maturity, since the latter can be accomplished through feedback of experience bearers as part of the collection process. A possible process to set up a measurement of EE time would be:

1. For each EE step estimate the time it needs to be performed.

2. For each maturity level, identify which tasks have to be performed to which extent (factor).

3. Sum up the time values.

The calculation of the EE effort highly depends on various unratable technical, organizational and human factors: specific environment, possible tool support for EE tasks, expertise and laziness of the experience engineer(s), formality level of the experience representation and the amount of collected experiences. For this reason, this thesis uses a simplified ordinal metric for EE effort by just *counting the number of EE tasks* as defined in Section 2.1.3 that have to be performed.

The amount of experience engineering that is needed before the EM initiative renders a benefit to the experience bearers, is defined as *minimal experience engineering effort*:

Definition 5.6 (Minimal experience engineering effort).

The minimal experience engineering effort is the experience engineering effort that is needed to achieve the desired optimal immediate benefit (see Definition 5.7) for experience bearers.

A *desired optimal immediate* benefit is the kind of benefit that can be immediately (see Definition 5.7) expected from an experience management initiative to provide noticeable improvement and help for experience bearers. This benefit, though, can be lower than the benefit of fully engineered experiences.

An example of minimal experience engineering effort would be to filter and anonymize experiences in order to share them with other partner teams. An overview of the different stages of benefit in relation to the experience engineering effort is presented in Section 5.5, Figure 5.4.

5.4.2. Administration Effort

Administration effort covers the technical aspects of experience management like setting up the environment, preparing and conducting the collection event. These tasks are not part of the experience engineering process according to Section 2.1.3. *Administration* includes the following tasks:

1. Installation of technical environment (e.g. experience base, plugins, recording systems, databases),

2. Configuration of technical environment (e.g. creating needed templates, rules, categories in the experience base),

3. Preparation for elicitation (e.g. preparing questions, scheduling),

4. Carrying out (moderation or attendance) and taking minutes during elicitation sessions.

5. Upload or save collected material. This can include transformation of collected data into another form that can be saved in the experience base, e.g. transcribing or scanning paper notes, video or audio conversion, interlinking documents with other artifacts.

All tasks, except saving and uploading, have a fixed duration. They have to be done in the same extent, independent of the number of collected experiences.

Like the EE effort, measurement of administration effort is highly contextual and requires knowledge outside of the collection method's process. Administration effort depends on the implementation of the collection method, the technical environment, the expertise of the administrator and existing rules or security policies. Even if administrative tasks and their effort are mentioned in the method's description in the literature, this may be transferrable to other systems, organizations or project settings only to a limited degree. Nevertheless, it is possible to estimate which administrative tasks are needed to be executed. This can give a rough indication of administration effort.

5.5. Measurement of Benefit

This section approaches the question how to measure benefit from an EM initiative. Benefit can be created by providing a helpful or useful EM initiative. Lack of perceived usefulness of the base's content is listed as a major experience sharing barrier (the barrier *Content perceived not useful* in Figure 3.1).

Since the focus of this thesis is to examine effort of an EM initiative from different perspectives, benefit must also be differentiated. According to the identified sharing barriers, perceived immediate personal benefit is crucial for the success of an EM initiative besides a helpful experience base. Assuming an experience management structure similar to the Experience Factory [30], a full-scale experience engineering process may be time-consuming and will not provide benefit immediately [33]. In a distributed collaboration, however, sharing policies may overrule the intent to share experience. Competitive relationship of partners and NDA agreements is a grave sharing impediment (barriers 1 and 2 in Appendix A). Due to this fact, immediate benefit is often not possible without any engineering steps. Raw experiences should be filtered or at least anonymized before sharing them with distributed partners. Thus, the (immediate) benefit should be regarded as a benefit with *minimal* experience engineering as defined in Definition 5.6, if the experiences are intended to be shared with dispersed partners.

On the other hand, it is also important to assess the long-term benefit the organizations and individuals would gain after the collected raw experiences have been fully engineered.

The following sections present measurements for both kinds of benefit: immediate benefit with minimal EE and benefit with full-scale EE. Figure 5.4 illustrates where to expect the benefit with different stages of EE.

Table 5.14 provides an overview of the possible benefits an EM initiative can produce and puts the benefits to perspectives they apply. It is partly based on Bergmann's work [36, pp.8-9].

Table 5.14.: Possible benefits from an EM initiative for each perspective (based on [36, pp.8-9]).

Perspective	Benefit
personal	- save time for task or process - lose dependancy from (asking) distributed team members - a task or process has become possible - reduce error rate for task or process
team	- save time for task or process - loose dependancy from (asking) distributed team - a task or process has become possible - reduce error rate for task or process
organizational	- processes have become more efficient - competitive advantage - cost reduction - project time reduction
inter-organizational	- processes become more efficient - gain knowledge

A possible way to measure benefit from the base's content would be to examine to what extent the EM initiative helps to improve work processes. More specifically, process improvement means that a process is completed or conducted more efficiently and with less errors. Efficiency is the amount of outcome in a given time. Thus, raised efficiency would mean saved time for a given task or process.

A further benefit of externalized experiences (on personal and team level) is to lose the necessity of asking colleagues in case of a problem. Low awareness and less effective communication are major problems in distributed collaborative projects (see Section 3). Unlike small co-located projects, where asking a colleague next door may be much more efficient than look up in the experience base, this does usually not apply in distributed projects [120, 119]. Thus, providing a source of shared and searchable experience can save time for dispersed team members.

5.5.1. Perceived Benefit with Minimal Experience Engineering Effort

This section considers perceived benefit of participants in a distributed software engineering project without or with just the minimal experience engineering effort (see Definition 5.6). Hereby benefit is measured, when raw experiences are directly extracted or elicited and not yet fully engineered into recommendations. I consider automatic processing steps of collected experience, which are

- restructure experience artifact by e.g. transformation into a new format, automatic anonymization, automatic tagging, clustering etc.,

- save experience artifact in the experience base,

but no manual steps that require an experience engineer after the experience has been submitted (see Figure 5.4). However, some collection methods result in experiences that may be too sensible or raw to be allowed to share with other project partner teams. Therefore, a minimal experience engineering process may be needed as well beside the automatic processing. In case the procedures needed to ensure information security (anonymization, filtering) can also be accomplished automatically, the here regarded benefit would be without any experience engineering.

It is also important to keep in mind that according to the sharing barriers, this kind of benefit is relevant on personal level in the first place and not on the organizational.

This thesis proposes a process to assess possible benefit with minimal or no EE:

1. Identify the potential learners and providers. Both should benefit from the collected experiences.

2. Estimate the time a learner and provider need to accomplish a task *without* the help of the experience base and appertaining tooling. The task must be central to the one supported by the EM initiative.

3. Estimate the time needed to accomplish the same task *with* the help of the experience base and appertaining tooling.

4. Estimate rawness and maturity of raw experiences collected by the experience collection method.

5. Calculate the saved time according to equation 5.4.

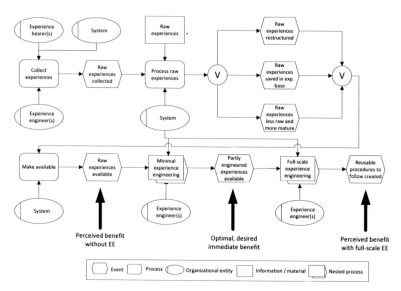

Figure 5.4.: Experience collection process marking when to expect possible benefit with full-scale, minimum and without EE. This diagram uses the Extended Event-Driven Process Chain (eEPC) [146] notation.

Expressed as a formula, saved time t_{saved} can be measured as:

$$t_{saved} = (t_{old} - t_{new}) \cdot \frac{\alpha \cdot m}{\beta \cdot r}, \tag{5.4}$$

where

t_{old} is the time needed to accomplish a task or process *without* the help of the experience base and appertaining tooling. The task or process must be central to the one supported by the EM initiative.

t_{new} is the time needed to accomplish the same task or process *with* the help of the experience base and appertaining tooling.

$\alpha \cdot m$ is the experience maturity with a factor α of raw experiences that have been already collected and are available in the experience base.

$\beta \cdot r$ is the experience rawness with a factor β of raw experiences that have been already collected and are available in the experience base.

Maturity m and rawness r have an anti-correlating relationship: A high maturity and low rawness of experiences should have a positive influence on the benefit beside mere time saving. The factors α and β could not be determined within the scope of this thesis. To compare the collected experiences and benefit, I assume an "optimal" experience base, i.e. a base without initial seeding problems [96, 213] and having a content that is theoretically helpful for the examined task. Here, I do not take into regard any active dissemination activities[5] and assume that the learners and providers are aware of the experiences available.

Immediate Benefit

EM initiatives should strive to render an immediate benefit, i.e. a benefit or reward perceived "within a reasonable time frame" [212]. In this thesis, immediate benefit is defined as:

Definition 5.7 (Immediate benefit).
An immediate benefit from an EM initiative is a benefit an experience provider perceives within one week after he has shared his experiences.

Personal and immediate benefit can also be produced through special reward mechanisms. This is discussed in more detail as related research in Section 7.4, though.

---

[5]Dissemination strategies are out of scope of this thesis.

5.5.2. Possible Benefit with Experience Engineering

This kind of possible benefit can be expected after collected raw experiences have been engineered e.g. according to the process in Section 2.1.3. This benefit is especially important for management of the collaborative partners, but also teams and team members would directly or indirectly profit from process improvement strategies. To create these assets is the overall goal of an EM initiative. Of course, the possible benefit can vary depending on the engineering steps that were conducted.

Although providing a benefit of engineered experiences is important to justify the existence of an EM initiative, this thesis focuses on possible immediate benefit with minimal or no EE. For this reason, this type of benefit will not be further examined.

5.6. Scale for Light-weight Experience Collection

In this section I derive a scale to estimate light- and heavy-weightedness of an experience collection method. After applying the measurement system defined in Sections 5.3 to 5.5 on the collection method catalogue from Section 4.4, I present them in a portfolio.

5.6.1. Methodology

In order to establish a scale of light- weightedness based on the catalogue of light-weight collection methods from Section 4.4, they must be contrasted to heavy-weight methods. Therefore, I conducted a literature review. The review methodology was similar to the one for light-weight experience collection methods in Section 4.4.1. The search string consisted of parts displayed in Table 5.15.

Table 5.15.: Keywords for the search.

| A1—knowledge | B1—heavy-weight | C1—collection |
| A2—experience | B2—heavyweight | C2—elicitation |

Like for the review of light-weight experience collection methods, the overall search strings were the elements of the power set of the key words in Table 5.15. The most results, however, resulted from a recursive search in the literature on light-weight experience collection methods or on effort as an experience sharing barrier [87, 215, 170, 212, 185, 210]. The inclusion criteria of a method as heavy-weight was, if the source mentioned this method as

- more effortful than the presented light-weight solution (e.g. mentally exhausting, lack of acceptance due to effort excess),

- explicitly described the method as "heavy-weight" or "heavyweight".

Appendix B lists all identified heavy-weight experience collection methods and also discusses the rationale of this assessment, including sources that claim why this method is heavy-weight.

5.6.2. Portfolio of Experience Collection Methods

According to Definition 4.3, a light-weight experience collection method must fulfill at least one of the five criteria (rephrased as a guideline): C1: extend or reuse well known tools, C2: integrate into existing tooling and processes, C3: keep time to share low, C4: minimize cognitive load and C5: delegate experience engineering (EE) tasks from experience bearers to others. The five criteria can be reflected by two factors: *time to share* and *cognitive load* (see Section 5.2). These two factors represent the axes in the portfolio. Sharing time directly covers the criterion C3 from the definition and indirectly C5 (shift experience engineering away), since delegating EE tasks from experience bearers lowers their sharing time as well. Criteria C1, C2 and C4 can be consolidated into the cognitive load score as defined in Section 5.3.4. The cognitive load consists score contains statements about process and tool adaptation or reuse as well as mental demand, which should be kept low in order to make the sharing process less exhausting. Organizational effort and benefit are not reflected in the portfolio, as they are not directly part of the light-weightedness criteria in Definition 4.3. Nevertheless, to provide a full assessment on effort as well as immediate benefit (on an organizational and personal level) of light-weight and heay-weight experience collection, the results of this assessment are presented in Section 5.7.

The portfolio is displayed in Figure 5.5. It shows the results of classifying and examining the methods from Section 4.4. The axes reflect both personal effort factors, the *time to share* and *cognitive load*.

The circles represent the methods, which are encoded according to Table 5.16 for the sake of a clear overview. The calculation of time to share and cognitive load, as well as the rationale behind the measurement results for the methods can be looked up in Appendices C and D. Note that the *Time to share* axis is logarithmic. This is due to the three top methods (HW1, HW2 and HA3), which have a overproportionally high time to share that would not fit into a linear scaled portfolio of reasonable size.

Unlike the usual procedure to classify items in a 3x3 portfolio (e.g. risks in a risk portfolio) and to use a gradient of $45°$, this portfolio has a rather flat limit gradient. Due to the logarithmic scale of the *Time to share* axis, a steep gradient would drastically increase the time to share for methods with a low cognitive load. On the one hand, to be classified as light-weight, I assume that a method with low cognitive load should be "rewarded" with a higher time to share (an easy but longer collection session) and a method with higher cognitive load should be "punished" with

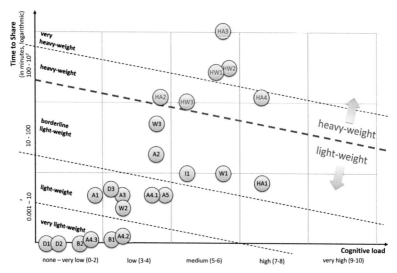

Figure 5.5.: Portfolio of all examined experience collection methods. For the codes refer to Table 5.16. The rationale on the computation of time to share can be found in Appendix C and for cognitive load in Appendix D.

less allowed time to share respectively (a more exhaustive session should be shorter). However, a steep, e.g. 45°, gradient would mean that light-weight methods with a low cognitive load could take months to share an experience. The gradient reflects a priority of sharing time to cognitive load. According to the experience sharing barriers (see Chapter 3), *effort > benefit* is the central obstacle to sharing experiences. The time to share should not rise too much even for methods with a low cognitive load.

The bolder dashed (red) line marks a limit between the heavy-weight and light-weight region. The finer black dashed lines mark finer gradations. All methods above the limit are the identified as heavy-weight methods. Here, they are coded with H*Xi*.

The methods HA1 and HA2 are not a result of a literature search but they represent possible counterparts to the light-weight authoring, by-product and data mining methods. HA1 (answer each question as expert without Answer Garden) especially should be the complement of Answer Garden. HA2 is the heavy- or rather heav*ier*-weight counterpart of the light-weight experience authoring methods that use a Wiki (e.g. *Wiquila*, A5). Wikis are a popular way to support collaboration and are often relatively easy to set up and manage [161, 34]. Wiki as an experience base

Table 5.16.: Codes for light-weight collection methods.

Category	Method	Code
by-product	CRMT	B1
	FOCUS	B2
data mining	e^3	D1
	Autom. Rationale Extr.	D2
	Delta Analysis	D3
interview	Reflective Guides	I1
exp. workshop	eWorkshop	W1
	LIDs	W2
	LPMR	W3
	Project History Day	HW1
	Learning Histories	HW2
	Post Project Review	HW3
exp. authoring	Observation Sheet	A1
	Answer Garden	A2
	Dandelion	A3
	Mail2Wiki A (part of email with plugin)	A4.1
	Mail2Wiki B (batch of emails without plugin)	A4.2
	Mail2Wiki C (batch of emails with plugin)	A4.3
	Wiquila	A5
	Write experience in Wiki	HA1
	Answer each question directly (without Answer Garden)	HA2
	Postmortem report	HA3
	Quality Patterns	HA4

and experience collection tool has nowadays become a popular strategy. However, according to own experience in a distributed collaborative project [179] and many other reports and numerous case studies (e.g. [252, 164, 80, 43, 128]) Wikis are often not well accepted and require effort and cognitive load to use. The goal to include this method into the analysis is to examine, if a Wiki is a light-weight method.

The placement of the light-weightedness limits is an estimation based on reasoning that the methods above the limit are assumed as heavy-weight. The definition and placement of the finer gradations *very light-weight, light-weight, borderline light-weight, heavy-weight, very heavy-weight* stems from the observation that the methods are roughly clustered. These clusters, especially the *very light-weight cluster*, are not only grouped according to their position but also according to the collection method categories and technical realization.

The identified clusters are:

- *Very light-weight cluster:* This cluster contains methods where the experience collection is most "automatized". These are the data mining methods D*i* and the experience as a by-product methods B*i*. The two experience authoring methods A4.1 and A4.3 (Mail2Wiki

A and B) only require forwarding a batch of emails or dragging and dropping them into an email plugin. It is not surprising that these methods have the least time to share and cognitive load for experience bearers. All of the methods only need a couple of clicks to start an extraction. Delta Analysis (D3), though a data mining method, is an exception to this reasoning and is only light-weight. This is owed to the focused interviews as part of the method, which requires effort from the experience bearers.

- *Light-weight cluster:* This cluster contains experience authoring methods Ai that allow ad hoc sharing of short experiences.

- *Borderline light-weight cluster:* This cluster mainly contains experience workshop Wi, and interview methods Ii as well as more effortful experience authoring methods.

The heavy-weight methods also slightly form two clusters:

- *Heavy-weight cluster:* The heavy-weight area contains moderately heavy-weight methods that require a couple of hours to share one experience and require a medium or high cognitive load.

- *Very heavy-weight cluster:* This group contains the most effortful and exhaustive collection methods ranging from several hours to months.

For easier classification, Figure 5.6 provides a simplified rough approximation for the five areas. To minimize the error in the approximation, the values are calculated as the mean time to share of the line inside a cognitive load quadrant.

Figure 5.7 presents the light-weight experience collection methods analyzed according to the number of fulfilled criteria from Definition 4.3 on page 35. The information about the fulfilled criteria of a collection method is taken from the light-weightedness rationale in the introduction of the method in Section 4.4. The distribution shows the highest density of methods with 4 and 5 fulfilled criteria in the *very light-weight* range. The *light-weight* range also only contains methods with more than 3 fulfilled criteria. The *borderline light-weight* region predominantly contains methods with less (2 or 1) fulfilled criteria. This analysis supports the assumption in Section 4.1 that the more criteria from the definition are fulfilled, the more light-weight the method is. Regarding this distribution, a slightly different segmentation of the light-weightedness ranges could also be justified. The *very light-weight* and *light-weight* ranges could be consolidated into one *light-weight* region, as they together contain 70% of methods with 4 and 5 fulfilled criteria.

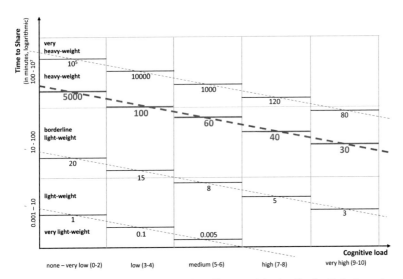

Figure 5.6.: Approximation of the light-weight and heavy-weight classification limits for easier measurement.

5.7. Lowered Experience and Knowledge Sharing Barriers

This section approaches the research question **RQ4**:

RQ4: *Do light-weight experience collection methods lower the significant experience and knowledge sharing barriers and increase participation in an experience management initiative in a distributed software engineering project?*

I discuss which sharing barriers can also be overcome with light-weight collection approaches beside the ones described in Section 4.2. This discussion is based on results from the assessment of light-weight method catalogue according to the defined effort and benefit measures. This section also compares heavy-weight and light-weight methods regarding these barriers.

5.7.1. Immediate Benefit

The analysis showed that all examined light-weight experience collection methods returned an immediate benefit to the experience bearers, while almost all heavy-weight sharing methods returned no or very limited immediate benefit. Figure 5.8 presents an extended version of Figure

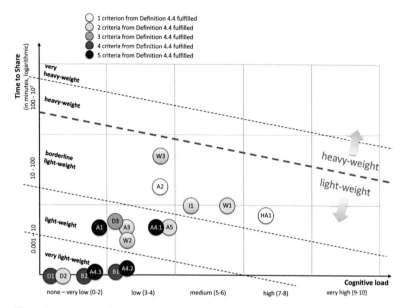

Figure 5.7.: Portfolio of all examined experience collection methods colored according to the number of fulfilled criteria from Definition 4.3 (page 35). For the codes refer to Table 5.16.

4.2 with more lowered barriers beside *effort > benefit* reflecting this insight. Tables 5.17 and 5.18 list the immediate benefit of light-weight and heavy-weight methods.

The most frequent benefit is time saving by reducing the necessity to ask distributed colleagues. For all heavy-weight methods except Quality Patterns (HA4), a long report needs to be written, which presumably takes longer than a week. Although literature about experience workshop methods like Learning Histories (HW2) or Post Project Review (HW3) contains statements that participants gain more insight or "Aha!"- effects during socialization with others, other teams (other learners) do not learn about this workshop immediately and thus do not immediately benefit from it. On the other hand, light-weight experience workshop methods like LIDs (W3) also provide insights through discussion and a motivating wrap up of events and immediately deliver an experience artifact, which is valuable and can be immediately accessible to others.

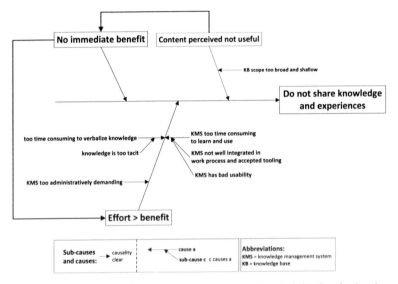

Figure 5.8.: Excerpt from the sharing barriers overview (Figure 3.1) depicting those barriers that are lowered by light-weight experience collection methods based on the assessment of the methods.

Table 5.17.: Immediate benefit (see Definition 5.7) of *light-weight* collection methods for a supported task, target user group and artifacts, which provide the perceived benefit. Refer to Table 5.16 for method codes.

Method	Perspective	Process or task to support	Immediate benefit without EE	Immediately available artifact
B1	risk assessment meeting participant	remember meeting and course of discussion without taking notes [215]	save time making notes or asking colleagues	record of chat linked to risk placement
B2	developer not present at prototype presentation	learn about the prototype [215]	save time making notes or asking colleagues	slides linked to speech and code
D1	any developers	search experts on specific topic	save time asking colleagues	expertise graph
D2	developer making decision, who has not participated in the meeting	search for rationale to a decision, prepare for next meeting	save time asking colleagues	rationale from meeting minutes as free text
D3	any developer	create and edit a project related document	save time searching errors, lower error rate and time to redo	deltas (differences in documents)
I1	developer participating in reflective guides	a) carry out e.g. requirement engineering activity (interview); b) gain knowledge on how to improve requirements engineering activity	save time reading books or ask colleagues; possibly reduce error rate during activity	reflection during answer of questions, socialization during workshop
W1	any developer	find experiences on e.g. defect reduction	save time trying; socialization with other experts during workshop	chat and report
W2	a) other development team; b) participants	ab) learn about experiences of the other team in specific project settings	a) save time asking; enable recollection of events that may not be remembered later; b) wrap up of events and saving them is motivating [212] wrap up of events and saving them is motivating [212]	LID document
W3	development team	learn relations of problems during project	possibility to learn relations of problems and avoid problems in the next project; socialization with other experts during workshop is "inspiring" for participants [87]	post-mortem report with root cause diagrams and prioritized issue list
A1	any developer	no specific process	save time asking colleagues; experiences are authentic due to emotion component	short experiences on a sheet or scanned in e.g. a Wiki [23]
A2	expert	provide expertise, answer questions	save time not answering each question personally	-
A3	any developers	no specific task or process	save time asking colleagues; reduce coordinative emails and save time writing and reading them	experiences in a Wiki
A4	any developers	no specific task or process	save time asking colleagues	experiences in a Wiki
A5	any developers	no specific task or process	save time asking colleagues	experiences in a Wiki
HA1	any developer	no specific task or process	save time asking colleagues	experiences in a Wiki

Table 5.18.: Immediate benefit (see Definition 5.7) of *heavy-weight* collection methods for a supported task and user group, as well as artifacts that provide the perceived benefit. Refer to table 5.16 for method codes.

Method	Perspective	Process or task to support	Immediate benefit without EE	Immediately available artifact
HW1	any developers	no specific task or process	deeper insight and make or name problems explicitly ("experience a gratifying 'Aha!'" [62])	none
HW2	a) participants	a) no specific task or process	a) venting ("opportunity to ruminate at work, at length, about his or her experience." [204])	a) none
	b) other teams	b) no specific task or process	b) no immediate benefit	b) none
HW3	a) review participants	a) no specific task or process	a) only socialization during review	a) none
	b) other teams	b) no specific task or process	b) no benefit, since other teams do not receive the report [155]	b) none
HA2	expert	provide expertise, answer questions	no benefit	none
HA3 (writing report)	any developer	no specific process	no immediate benefit for team members	none
HA3 (post-mortem meetings)	any developer	no specific task or process	rebalanced schedule	none
HA4	any developer	no specific task or process	save time asking colleagues	experiences with quality patterns

Table 5.17 also shows that the majority of the light-weight methods support a quite specific process or task. Thus, these methods have a narrow focus, which is desirable for a successful experience base [214, 220]. This lowers the barrier *KB scope too broad and shallow* in Figure 5.8. Almost none of the examined heavy-weight methods support a specific task.

5.7.2. Organizational Cost

Organizational cost is reflected through effort and man hours an experience engineer spends on experience engineering and administrative tasks for the EM initiative. One would expect a light-weight method to be very costly to engineer and administrate according to C5 (shift experience engineering away) in Definition 4.3. Considering the *minimum EE effort* which is needed to return the immediate benefit, light-weight methods do not require much more EE effort. Table 5.19 depicts the minimal EE effort and administrative effort for all evaluated methods, while Appendix E provides the rationale on the estimated values.

Most of the experience authoring methods do not require any EE effort besides anonymizing, rephrasing and filtering raw experiences. In a distributed collaborative setting these tasks are often necessary to ensure that no confidential information can pass to parties who have no rights to access it due to NDA policies. Some methods do not require any EE activities, not even anonymization and filtering: *FOCUS* (B2), *CRMT* (B1) and e^3 (D1)[6]. For these methods it can be assumed that the raw experiences are shared in a meeting with partner teams or with the knowledge in mind that the experiences will be read by other teams. I assume that these experiences will a priori not contain sensitive information. The experiences collected with *FOCUS*, for example, would come from an official presentation, where only non-classified information is shared with partners. As an exception, *eWorkshop* (W1) and *Reflective Guides* (I1) need substantial experience engineering effort, since the experience engineer has to create reports. Here, I do not consider different lengths, analytical depths (just summarize or analyze and conclude) or specific rules how these reports have to be created.

Some heavy-weight collection methods also do not require any EE effort, e.g. *Postmortem Report* and *Answering each question without Answer Garden* (A2). These are the methods, where all EE tasks are done by the experience bearers themselves. Most heavy-weight methods require the full scope of EE tasks. Combined with enormous time costs from the experience bearers' side, they pose a very high organizational cost.

The *administrative effort* for light-weight collection methods, assuming that the needed tool support does not have to be developed, is also not much higher than the one for the heavy-weight methods. As a conclusion from this analysis, light-weight methods require on average much less

---

[6]These methods are introduced in Section 4.4.

Table 5.19.: Organizational effort with task IDs to be performed. Experience engineering tasks are presented in in Section 2.1.3 and administrative tasks are displayed in Section 5.4.2. The gray shaded methods are classified as heavy-weight.

Category	Code	Method	Min. EE effort	Admin. effort
by-product	B1	CRMT	none	1, 5
	B2	FOCUS	none	1, 5
data mining	D1	e^3	none	1
	D2	Autom. Rationale Extr.	1, 2	1
	D3	Delta Analysis	1, 2	3, 4, 5
interview	I1	Reflective Guides	1, 3, 4, 6, 8	3, 4
exp. workshop	W1	eWorkshop	1 - 8	1, 2, 3, 4
	W2	LIDs	3	3, 4, 5
	W3	LPMR	1 - 8	3, 4, 5
	HW1	Project History Day	1 - 8	3, 4, 5
	HW2	Learning Histories	1 - 8	3, 4, 5
	HW3	Post Project Review	1 - 8	3, 4, 5
exp. authoring	A1	Answer Garden	3	1, 2
	A2	Dandelion	3	1, 2
	A3	Observation Sheet	3	1, 2, 5
	A4.1	Mail2Wiki A (part of email with plugin)	3	1, 2
	A4.2	Mail2Wiki B (batch of emails without plugin)	3	1, 2
	A4.3	Mail2Wiki C (batch of emails with plugin)	3	1, 2
	A4	Wiquila	3	1, 2
	HA1	Answer each question directly (without *Answer Garden*)	3	1, 2
	HA2	Write experience into a Wiki	3	1, 2
	HA3	Postmortem Report	none	none
	HA4	Quality Patterns	3	1, 2

minimal EE effort and not more administration effort to render a benefit to experience bearers. Theoretically, as they render immediate benefit, they are not required to conduct the whole EE process to be considered useful and encourage participation and adoption. Heavy-weight methods, on the other hand, specifically the experience workshop methods, must conduct the whole EE process before producing any benefit. On the other hand (and in accordance with Definition 4.2), heavy-weight methods promise higher quality results than light-weight. This is reflected in the maturity and rawness of analysis of collected experience, which is presented in the following subsection.

Maturity and Rawness

Considering maturity and rawness of collected experiences, heavy-weight methods help to collect experience in a more mature and refined state. The portfolio in Figure 5.9 (or see also

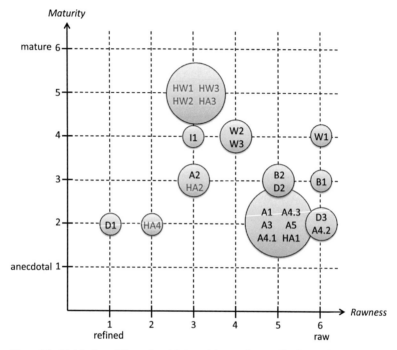

Figure 5.9.: Distribution of the analyzed light-weight experience collection methods according to their maturity and rawness. For the codes, refer to Table 5.16 on page 101. The red colored methods are heavyweight according to Figure 5.5.

Appendix F) shows a distinctly higher maturity and lower rawness for heavy-weight methods than for light-weight ones. The higher maturity can be explained by the fact that almost all heavy-weight collection methods are (post-mortem) experience workshops. The collected experiences there discussed with the whole team or even teams. This would also be the case for light-weight experience workshops. The lower rawness of heavy-weight methods can be explained by not abiding by C5 (shift experience engineering away) of light-weight experience collection (see Definition 4.3). Many EE tasks are integrating into the collection process, producing experience of higher refinement. However, this drawback of light-weight methods can be mitigated by the fact that these methods deliver an immediate benefit. An EM initiative can still be successful delivering experiences with a lower maturity and higher rawness but *immediately*, than an initiative that does not render any immediate benefit on the personal level at all [212].

As a summary, Table 5.20 compares light-weight and heavy-weight methods by displaying mean values and approximations of the measured effort, benefit, maturity and rawness factors.

Table 5.20.: A general comparison between light-weight and heavy-weight experience collection methods according to their average, effort, benefit, maturity and rawness.

	Light-weight		Heavy-weight
time to share	7.0min	$\longleftrightarrow$	$340999.3min \approx 236.8$ days
cognitve load	3.2 (low)	$\longleftrightarrow$	4.8 (medium)
immediate benefit	yes	$\longleftrightarrow$	no
minimal EE effort	low	$\longleftrightarrow$	high
administration effort	medium	$\longleftrightarrow$	medium
maturity	2.6	$\longleftrightarrow$	4.2
rawness	2.3	$\longleftrightarrow$	4.2

5.8. Discussion

The field of experience collection is fuzzy, as personal effort and benefit involves people and their attitudes. This thesis provides an important first step towards objective measurements for experience collection effort and benefit. Objective measurements can only be applied on unvarying subjects and must return the same result if repeatedly applied. A typical example of an objective measure is Lines of Code (LOC). Subjective measures, on the other hand, depend on the measured objects and the viewpoint from which measure is taken. The results achieved with this measure can differ with each repeated measurement of the same object, as people and their attitudes vary and change. Measuring skill and usability are typical subjective measures [258].

The classification of the light-weight methods and subsequently the construction of the light-weight experience collection portfolio (Figures 5.5 and 5.6 in Section 5.6) base on specific implementations of the methods as published in the literature. Their validity and generalizability may be limited. The measurement results could have been different if the researchers or practitioners had employed the collection methods differently. For instance, *eWorkshop* would be less lightweight, if it was held more than 3 times during a 2-years period.

Beside the predictable and objectively measurable factors, there are a number of additional subjective factors that influence experience sharing effort and the benefit. These factors were not taken into account. They are either not predictable depending on the specific situation or setting, or could not be set without limiting the applicability of the collection method. This fact has implications for the transferability and generalizability of the concepts and measurements: The measurements should be only considered as indicators and approximations. For each examined method, the measured values and the rationale on how they were derived are documented in Appendices B-F.

Table 5.21 and the following description provide an overview and details on factors that also influence effort and benefit and which were *not* considered in this thesis.

Table 5.21.: Factors that also have influence on the examined factors for effort and benefit but were or could *not* be measured in this thesis.

Main factors	Examined factors	Not measured factors
Effort	Experience engineering (EE)	- Maturity and rawness relation - Personal skill of experience engineer
	Administration	- Implementation of tooling - Development need - Personal expertise - Technical environment - Security policies
	Time to Share	- Personal (language and work) skill - Tool usability and accessibility
	Cognitive load	- Personal reading, writing and language skill - Tool usability and accessibility
Benefit	Immediate benefit (possible benefit with minimal or no EE)	- Usefulness of support for concrete task or problem - Dissemination strategy, tool usability and accessibility - Maturity and rawness relation
	Possible benefit with full-scale EE	- Specific company goals - Collaboration and knowledge sharing policies - Experience engineering process steps

Experience engineering (EE) effort It is reasonable to assume that experience maturity and rawness have an influence on the EE effort. A high maturity would mean that less effort has to be invested into step 9 of the EE process (refer to Section 2.1.3). A low rawness would mean less work for steps 2–7. However, it is not clear how *exactly* maturity and rawness influence these tasks. Personal skill and domain knowledge of the experience engineer can also influence the EE effort.

Administration effort The unknown variables for administration effort are technical environment and tooling as well as policies and personal expertise. Also cost and time to develop the tool support or establish the collection process were not taken into consideration. The results in Table 5.19 only include the effort for administering an already existing tool support and processes.

Time to share and cognitive load Not considered factors for sharing time are personal language skill and the ability to work efficiently. Some people may need longer to accomplish a task than others. Also the usability and accessibility of a tool can play a role in the time to share but also for the cognitive load. The tool support may e.g. be highly integrated into present tooling, but be hidden in a sub-menu of the main program and not support the main work tasks well. Thus, it would be badly usable and accessible, raising the experience sharing time and cognitive load. Due to a lack of sufficient information in the publications about the subjective factors, I assumed average values as default (see Tables 5.2 and 5.3 in Section 5.3) to calculate time to share. Having chosen other default values, the light-weight portfolio could have looked differently. Thus, if classifying a new collection method and comparing it with the other methods analyzed in this thesis, the default values in this thesis should be used if no information is present.

Immediate benefit The indicator for benefit through saved time (Equation 5.4) calculates if a person has a time saving doing a task (supported by the EM initiative) with and without the help of experiences and tooling provided by the EM initiative. This formula, however, does not consider if the supported task is needed to be supported in the first place, i.e. if the support is useful for the software engineer in his work. I assume that it is the case in general. All tasks and processes supported by the various light-weight collection methods presented in this thesis are directly related to SE development process or to general daily business activities. An experience dissemination strategy would also greatly influence the perceived benefit of an EM initiative. This wide research area is, however, out of scope of this thesis. As for the EE effort, the specific relation of maturity and rawness in Equation 5.4 has been yet unclear, except their anti-correlating nature.

Possible benefit with EE The possible benefit that engineered experience can return was not explicitly measured in this thesis as it is mainly organizational and not immediate. It depends on the company's specific goals, collaboration and sharing policies as well as the extent of the experience engineering effort.

Generally, using objective criteria in this thesis, the effort and benefit metrics do not measure the *perceived* but rather the *net* effort and benefit. Nevertheless, these estimations provide an indication with higher probability that a method with low net effort and high benefit will also be *perceived* this way. However, this cannot be guaranteed.

A further issue which was not discussed in this chapter is the question, if the constructed measurements are optimal or even sufficient to measure the specific effort and benefit aspects. Future work may approach this.

5.9. Contributions

This chapter presents the second part of the framework to assess light-weight experience collection. It answers **RQ3** by presenting a measurement system. This system allows to measure personal experience sharing effort by calculating the time to share an experience and cognitive load. For benefit calculation, this chapter differentiates between immediate and personal benefit (with minimal or no experience engineering effort) as well as organizational benefit with full-scale experience engineering. For a knowledge manager to employ a light-weight method, he must also be able to assess its organizational cost and benefit. As light-weight experience collection prescribes to defer experience engineering tasks from experience bearers, experience engineering and administrative tasks have to be carried out by the organization. The measurement system in this chapter provides measurement approaches for both types of organizational effort.

Applying the measurement system on the catalogue of light-weight experience collection methods from Section 4.4, this chapter further presents a portfolio of the analyzed methods and proposes a scale of light- and heavy-weightedness with altogether five gradations. From the assessment of the identified light-weight experience collection methods in Section 5.7, it can be concluded that light-weight collection methods help to overcome the effort to benefit disparity and the lack of (immediate) benefit.

6. Case Studies

This chapter approaches the main research question **RQ4**:

RQ4: *Do light-weight experience collection methods lower the significant experience and knowledge sharing barriers and increase participation in an experience management initiative in a distributed software engineering project?*

To investigate **RQ4**, I conceived two light-weight collection methods (Sections 6.1 and 6.2) for an experience management (EM) system. The system (Section 6.3) particularly supports global software engineering projects and defines an adapted experience engineering process and an experience base. The two collection methods are integrated into the system and fall into two experience collection method categories: data mining and experience authoring.

Section 6.1 describes a text analysis method named *Annotation Extractor (AnnoEx)* to extract annotations – and indirectly knowledge and experiences – from project documents to support collaborative research and document creation. In Section 6.2 I introduce an experience authoring method named *Observation Widget with Heuristic Support (OWHS)* to support experience bearers to write experiences of higher quality. The method AnnoEx mainly focuses on minimizing the extraction effort, OWHS focuses on low-effort capturing of ad hoc experiences that have a lower rawness. Figure 6.1 (on page 116) presents an extended collection method portfolio placing the evaluated collection methods and their variants. The classification rationale is described in the following sections in more detail.

Table 6.1.: Codes for evaluated methods in Chapter 6.

Category	Method	Code
text analysis	Annotation Extractor (AnnoEx) A	D4.1
	Annotation Extractor (AnnoEx) B	D4.2
exp. authoring	Observation Widget with Heuristic Support (OWHS)	A6

Section 6.3 describes the experience base and the experience engineering process for distributed software engineering, to present a possible (fictitious) EM system and how these two experience capturing methods interact with this system.

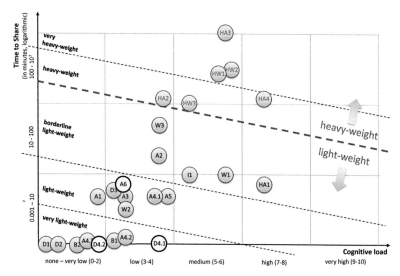

Figure 6.1.: Portfolio of all examined experience collection methods including those from the case study: AnnoEx A (D4.1), AnnoEx B (D4.2) and OWHS (A6). Refer to the codes in tables 5.16 and 6.1.

6.1. Annotation Extractor

In this section, I describe the main concepts and motivation behind the two variants A and B of the Annotation Extractor (AnnoEx). Then I demonstrate that this method can be classified as very light-weight and justify it. Afterwards, I examine the method's perceived usefulness with the help of a preliminary evaluation. At last, I examine whether AnnoEx requires much organizational effort. The description and evaluation of AnnoEx is based on Averbakh et al. [24].

6.1.1. Description

Many distributed collaborative software development projects follow structured and formal processes like Waterfall, V-Model or RUP. In such a process the team's work mostly revolves around documents that are created in a distributed fashion. Project participants often make annotations in these documents for different purposes, e.g. review or quick retrieval. The goal of the AnnoEx

116

[245] is to preserve knowledge and experiences that are contained in project document annotations. An annotation can be e.g. highlighted text, a note beside the text or graphic or any text corrections (see Figure 6.2). Annotations are created for different tasks in mind. These are collaborative writing, review, tagging important passages, or marking rationale behind a decision [167, 192].

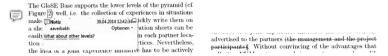

Figure 6.2.: Examples of different annotation types. Left: highlight and a textual note, right: a text correction (cross out). Source of annotated text: [23].

As a motivating use case, consider a situation, where a team in a distributed collaboration has to create a software artifact, e.g. a specification or a design document (see process in Figure 6.3). The process starts with a requirements engineer, who, together with a customer, writes a

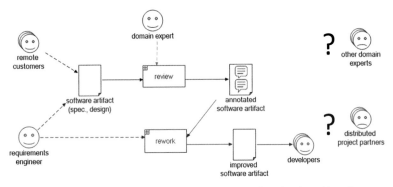

Figure 6.3.: Part of a (simplified) development process in a distributed setting without document annotation sharing.

specification. The output of the writing process is a software artifact (a draft) that requires a review. This draft is reviewed by a domain expert from the requirements engineer's company. The expert corrects the document by creating annotations, which are often based on experience. After receiving the annotated drafts from the reviewers, the requirements engineer reworks the draft and creates an improved version of the document, which serves as input for the developers. There can be, of course, more than one review cycle. In such a scenario usually only the final document is made accessible to others than the reviewers. The annotated documents are usually stored privately and are not accessible to, e.g. other domain experts or distributed partner teams. After the project, they are deleted and are not available any more.

However, annotations can contain valuable knowledge and experience. Example use cases could be:

Reveal possible flaws in a template or need for training: An analysis for clustering of specific types of annotations (e.g. annotations of unclarity or problems) could denote shortcomings in the provided template. It could also be an indication for lack of expertise and need for training in the particular area.

Retrieve valuable domain knowledge: Domain experts share their knowledge and experience when writing a review annotation. This valuable asset should be stored, engineered and shared.

Raise reviewing efficiency: In case of distributed collaborative reviewing of a document, it may be sometimes practical for a reviewer to know, which annotations partner reviewers have already made. This way he can adjust his review by, e.g., focusing on other issues to avoid double annotations or review other parts of the document.

Our approach is to store (potential) knowledge from annotations (see Figure 6.4) in a knowledge base, where they can be searched and commented. They can be engineered into recommendations in the long run.

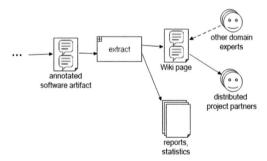

Figure 6.4.: Example scenario with AnnoEx as solution.

Besides annotations, AnnoEx also captures annotation metadata like author, date and annotation context (the marked and surrounding text or picture). The annotations and the original annotated document are displayed in a Wiki-based experience base allowing collaboration and search within the annotations. The experience base is introduced in Section 6.3.

Figure 6.5 presents an example Wiki page with extracted annotations. The annotations are ordered according to their vertical placement within a page in the document and grouped by appearance based on their page. The screenshot displays two extracted annotations from page 1

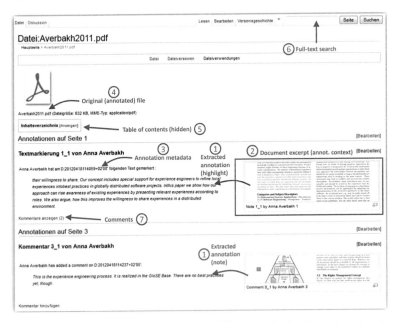

Figure 6.5.: Extracted annotation presentation as a Wiki page in the shared experience base (a Wiki).

and 3 one underneath the other. The annotations are separated by horizontal lines. The textual blocks at the left page side are the extracted content from the annotations ①. In the upper annotation the extracted text was highlighted and in the lower annotation it is a note. The annotation context is a screenshot of the annotation surrounding in the original document. It is placed at the right side next to the extracted annotation text ②. Above each extracted annotation, the page displays annotation metadata including author, creation timestamp and annotation type ③. Beside annotations, the reader can also access the annotated document, which is also uploaded to the Wiki ④. The reader has two ways to quickly scan annotation on the page. First, he can use a table of contents ⑤. This is hidden in the screenshot and has to be expanded. As an alternative, he can use the default full-text search of the Wiki, which would also scan the annotation text ⑥. The annotations can be further commented in the Wiki by other colleagues, thus activating and capturing new knowledge and experience ⑦.

AnnoEx falls in the category of text analysis, as it analyses the document structure to extract annotations. It is not an experience as a by-product method (see Section 4.3.1) as it does not capture the annotations in the process of their creation but afterwards. As a secondary method category, it supports distributed collaboration providing a shared platform and allows to comment annotations.

There are two variants of the AnnoEx concept (D4.1 and D4.2 in Figure 6.1):

AnnoEx A : In this version the experience bearer manually initiates annotation extraction from one or more documents in one extraction session. Thereby, he selects one or more annotated documents and sends them to AnnoEx in the documents' options menu.

AnnoEx B : The upload process is automatically initiated by a script. The experience bearer does not have to initiate the upload any more. AnnoEx would monitor one or several folders and process all documents in this folder.

AnnoEx A should be used in the case, where only some selected annotated documents should be uploaded. It provides a better way to keep awareness and control over the documents to upload. AnnoEx A can also be convenient in a situation, where the document author working with the annotated document, can upload it right after he finished annotating.

AnnoEx B can be utilized if a company (or collaboration) policy allows for uploading and sharing all annotated documents. They can be placed in one or more folders that are monitored and regularly uploaded. This AnnoEx variant reduces the personal effort of uploading to zero.

At the moment, only AnnoEx A is implemented, but it can quite easily be modified to AnnoEx B, e.g. by a script or scheduled task to monitor folders.

6.1.2. Light-weightedness Classification

According to Definition 4.3 all criteria C1–C5 are fulfilled. The method does not impose special software on the experience bearer (C1). The AnnoEx prototype is a Java (annotation extraction part) and .NET (writing annotations into Wiki) application in its current implementation and thus only works in a Microsoft Windows operating system (OS). Considering this fact, it could be objected that AnnoEx *does* impose a special software, namely the OS. However, the concept of AnnoEx is OS independent and can rather easily be ported to a different OS. AnnoEx extracts annotations from common text editors or viewers like Microsoft Word or PDFs. The annotation extraction works automatically unburdening the experience bearer (C3). It only needs one activation over a menu integrated in the system menu of the documents (C2). The goal of the integration is also to minimize cognitive load of a context switch, if the experience bearer would have to start a stand-alone tool to extract annotations (C4). C5 is also fulfilled, as the method does not require extra effort for the experience bearer after the extraction.

Time to share

To measure how light-weight AnnoEx is, its time to share must be calculated. AnnoEx extracts project document annotations into an experience base. Therefore, the variant AnnoEx A provides an option in the MS Windows system menu of the document(s) that triggers the annotation upload.

According to Formula 5.1, the overall time to share t_M for one experience during a 2 year-period with AnnoEx A is calculated as $t_M = \frac{3 \cdot 3}{229} \cdot (2\text{sec} + 4\text{sec} + 2\text{sec}) \approx 0.3\text{sec}$. The AnnoEx B configuration does not require any additional actions from the experience bearer making the time to share 0sec. Next, I explain my choice of values for the time to share calculation for AnnoEx A and B.

$t_{i,M}$: AnnoEx provides an option in the MS Windows system menu of the document(s) (Figure 6.6, ①) that triggers the annotation upload ② with only a couple of clicks. The duration values are the default click durations from Table 5.3. For AnnoEx B, there are no active

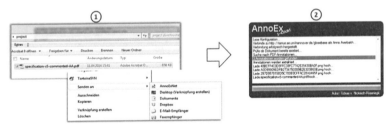

Figure 6.6.: GUI and interaction with AnnoEx.

steps for the experience bearer to perform. Here I assume, that he already has a special work folder, which can be monitored by AnnoEx.

f : *AnnoEx A:* I consider specification, design and user manual as documents that are likely to be reviewed and annotated because they contain free text. I assume that each of the reviewed documents have 3 review cycles. I further assume that the annotated documents are submitted after each creation of a final document version.

e : To determine how many experiences an annotated document can contain, I conducted an annotation analysis [24]. I analyzed 7 specifications from a student software projects course in 2013 / 2014. This course is part of the Computer Science Bachelor study program at the Leibniz Universität Hannover[1]. During the project, student groups of five have to develop a software which includes writing a specification. Each development group is assigned a customer and a coach. The customer provides the requirements and the coach supports the team. These two roles are usually taken by Software Engineering research assistants. The requirement documents are reviewed by the customer and the coach respectively. The customer usually corrects false or missing requirements. The coach tries to help the team by, e.g. marking dangerous passages (like too many guarantees in use cases), vague descriptions, wrong content according to the given template, and others. I only considered those documents that were reviewed by the team coaches for this analysis. I assumed that coach-reviewed specification would contain (more) experience-based recommendations which can be helpful across projects. To determine the percentage of annotations that contained an experience, I manually read each annotation and decided if it was an experience. An annotation was considered an experience, if it contained an *implicit or explicit recommendation*. Corrections about UML syntax was not considered experience, but knowledge (from books). The analysis revealed that about 60% of all annotations (178 out of 284) were experience-based. On average, each document contained about 25.4 experience annotations. Assuming 3 reviews for each document and assuming 3 documents to review (specification, design and user manual) this means 9 reviewed versions. This makes $9 \cdot 25.4 = 229$ annotations altogether.

p : I consider only one participant for each document, though more than one person reviewed the document and created the annotations. However, I do not consider the annotation authors in this equation, only the person who operates AnnoEx to upload the annotated documents. The reviewers would have the effort of annotating the documents anyway, independent of AnnoEx. Experience *sharing* happens by uploading the documents. I assume that the author of the reviewed document, i.e. the receiver of reviews, does the upload. I also assume that for each of the three documents there are different people in charge of the upload.

Summarizing, the time to share values are presented in Table 6.2.

---

[1]For a more detailed description of the software project course refer to e.g. Singer's dissertation [228].

Table 6.2.: Time to share t_M for AnnoEx.

Method M	Steps and $t_{i,M}$	f	e	p	t_M
AnnoEx A	(i) select documents with mouse: 2 sec.; (ii) open context menu and select "Send to An-noEx": 4sec.; (iii) confirm successful upload: 2sec.	3	229	3	0.3sec
AnnoEx B	no steps	nr	nr	nr	0sec

nr = not relevant

Cognitive Load

According to the measurement criteria in Section 5.3.4, the cognitive load for AnnoEx can be classified as low. Table 6.3 presents the cognitive load factors for each of the two AnnoEx variations.

Table 6.3.: Cognitive load for AnnoEx.

Method	Recall	Mental demand	Tool integr.	Process integr.	Ad hoc sharing	Cognitive load
AnnoEx A (manual upload activation)	1	0	0	1	2	4
AnnoEx B (folder monitoring)	0	0	0	0	2	2

Recall : *AnnoEx A*: The experience bearer must recall to select the chosen annotated files and also recall where to look for the menu to activate the document upload (in the document's menu). Thereby, I assume that his work process requires a predefined folder for annotated documents. *AnnoEx B*: The experience bearer does not have to recall any syntax, rules or processes to start the annotation extraction process.

Mental demand : *AnnoEx A, B*: The experience bearer has nothing to read or write and does not have to think analytically.

Integration into present tooling : *AnnoEx A, B*: Opening AnnoEx is well integrated into the operating system's native file context menu (see Figure 6.6). Thus it is fully integrated into the tooling environment.

Integration into work processes : *AnnoEx A*: The upload of a document is an extra task that has to be actively initiated and interrupts the work process. However, AnnoEx uses common practices and behavior by being well integrated into the operating system. *AnnoEx B*:

Since the experience bearer does not have to perform any additional actions, his work process is not interrupted.

Ability of ad hoc sharing : *AnnoEx A, B*: Experiences are already externalized (as annotations) at the time of sharing (uploading).

Benefit

We evaluated the usefulness of the annotation extraction and sharing concepts as a light-weight data mining instantiation in the software engineering domain. The evaluation and the results in this section are an excerpt from our work [24]. Merely the identifier names for the questions are changed. We identified the following four research questions (QI–QIV):

QI Is persisting and sharing annotations useful for software engineering tasks?

QII What kind of annotations are used in the software engineering field?

QIII Is our annotation extraction approach helpful? In particular, does our approach save time reading a document to find relevant information to solve a problem?

QIV Can annotations be used to improve software engineering processes? Specifically, can review annotations denoting frequent mistakes of a kind give evidence to flaws and unclarity in document templates?

A positive result for QI could give an indication towards a general usefulness of extracting and sharing annotations with other teams, while With QII and QIII we examine, whether AnnoEx returns an immediate benefit. A positive answer to QIV would also indicate a long-term benefit.

Methodology: We conducted three evaluations. To answer QI and QII, we distributed a survey (see Appendix H.2) to software engineers in three middle sized and big software companies. The software engineers were given questions about their annotation behavior and whether our approach could be helpful for their work. To answer QIII, we conducted an evaluation as a cross validation. Our participants were software engineering research associates. The evaluation was conducted at their workplace. The participants hat to fill out a questionnaire. The questions are displayed in Appendix H.1. We implemented an annotation extractor for PDF[2] documents that automatically uploads annotations into our Wiki-based experience base. We divided our evaluation participants into two groups: G1 and G2. The evaluation consisted of two parts: 1) the participant was given a PDF publication and had to answer several questions concerning the paper and 2) he had to answer questions concerning another paper with the help of our PDF annotation extraction prototype. The participants from the control group G2 had the same procedure but with switched publications. The publications were chosen to be tedious to read

---

[2]We have observed that researchers more often read and annotate PDF than MS Office documents.

without the possibility to quickly scan the content. They were long (23 and 31 pages), relatively unstructured with long textual blocks and little figures. Their topic was chosen not to be too far from the publications the participants would normally read (e.g. no chemical formulas) but also not too related to their research field. The questions were not directly searchable in the papers.

To answer QIV, we analyzed 32 annotated requirement specifications (12 PDF and 20 Word documents) from several reviews in the student software project mentioned above. This software project is mentioned in the rationale of the experience size e for time to share calculation. To create requirements specifications, students were provided templates containing a predefined structure with helpful questions and descriptions. Altogether, we extracted 697 review annotations, mostly in form of textual notes. To extract annotations denoting unclarity, we used a rather simple heuristic. We assumed that a search for annotations containing a "?" would yield questions, i.e. messages of unclarity. We identified 283 annotations containing a question. We assume that this way we can identify the most unclear chapters or places in the requirements template.

Results: Altogether, we received 6 survey responses from the industry. The participating software engineers consisted of 2 developers, 1 analyst, 2 architects and 1 project manager. Our results indicate that QI can be answered positively: 4 out of 6 participants consider annotating project documents very helpful and 1 rather helpful.

To answer QII we found out that software engineers mostly annotate for review purposes (with and without being involved in writing the reviewed document) and as a reminder to retrieve information quickly (5 answers). Annotating for the purpose of summarization or abbreviation of content as well as marking text to draw attention to it is less frequently used (2 answers). The most highly rated annotations to share were the reminders (3 answers). Sharing review annotations was favored by 2 participants. All participants annotate specifications, but rarely diagrams, presentations, publications and other project-related (e.g. mercantile) documents. The most often annotated file types were Microsoft Word and Excel (3 answers), but also a lot of printed documents. PDFs, PowerPoint documents and screenshots are less often annotated.

It is not very surprising that question QIII can be answered positively, indicating that annotations reduce the cognitive load when reading a long document. 6.7 shows that the overall 9 participants needed distinctly less time to answer the questions with annotated papers than without annotations: G1 needed 5.7min without and 2.5min with the help of annotations as median. For G2 the median was 8.0min and 3.0min.

After the evaluation we received feedback that most of the participants were frustrated and even upset because they took very long to read the papers. One participant even didn't answer the questions without the help of annotations, but wrote:

"Do you really expect me to read a paper with 23 pages?"

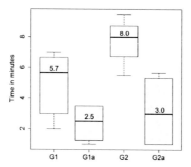

Figure 6.7.: Boxplot showing answering durations for both groups without (G1, G2) and with the help of annotations (G1a, G2a).

He answered the questions with the annotated paper, though. This indicates that annotations lower the barrier to read long documents if you lack the motivation. All in all, annotations and their presentation were perceived as helpful or rather helpful by 7 out of 9 academic participants. From the survey in the industry, 3 out of 6 participants stated that our annotation extraction concept could successfully support their work. Two participants did not find it useful (1 person abstained). One participant noted that searching annotations in the Wiki is helpful in a situation, where he does not remember in which document to look for.

To answer QIV, we found out that most annotations (74) denoting unclarity were on pages where the students had to write use cases and draw use case diagrams. This finding is supported by the fact that the most frequent term in all annotations of unclarity was "use case". This analysis indicates that these template chapters lack a good example or explanation. It could also mean that the students lack expertise about how to create good use cases. After project end, the students had a post-mortem session where they could share their feedback and experiences during the project. Many groups complained about their problems with use cases and lack of a good use case example in the template. This feedback also supports the results of the annotation analysis.

Organizational Cost

The organizational effort also lies within the limits of the other light-weight text mining method instantiations. Table 6.4 sums up the organization cost, maturity and rawness. The minimum experience engineering effort value indicated the EE tasks needed to perform (according to Section 5.4.1), administration effort denotes the administration tasks (Section 5.4.2), while maturity and rawness are scores as defined in Sections 5.3.2 and 5.3.3.

Table 6.4.: Organizational cost, maturity and rawness of AnnoEx.

Min. EE effort	Admin. effort	Maturity	Rawness
3	2	2	6

AnnoEx can also be used to share any documents, not only those that are explicitly meant to be seen by partner organizations. Therefore a filtering and anonymizing step must be performed as minimal experience engineering effort. Administratively, the experience engineer must configure the experience base Wiki, which I assume already exists.

The maturity of annotations is low (2), as they source from one reviewer at a time and are not confirmed or disapproved in the whole team. As a drawback of very low personal effort, the rawness of the extracted experiences is also very high (6). As mentioned in the rationale of time to share of AnnoEx, the annotation analysis revealed that about 60% of the annotations were classified as experiences. Many annotations were merely grammar or expression mistakes. This makes it more laborious for the experience engineer to filter out valuable experience. Of course, he could employ text mining methods to conduct a presorting. The annotation analysis showed that mining annotations can give indication about the quality of the template.

Summing up, AnnoEx can be classified as a light-weight collection method according to Definition 4.3 and the metrics defined in Chapter 5. An evaluation of AnnoEx's usefulness is an excerpt from our work [24] with changed research question names. It showed that annotation extraction and their presentation in the Wiki was perceived as immediately useful in the use case of searching for information in a long document. Engineering annotations, though having a very high rawness and low maturity, can be supported by text mining. The analysis showed this by indicating a template weakness. This can be regarded as an example of a long-term benefit of collecting project document annotations.

6.2. Observation Widget With Heuristic Support

Beside AnnoEx (see Section 6.1) the method in this section provides a possibility to explicitly share experiences through authoring. Like for AnnoEx, the method is described first, then showed why it can be classified as light-weight. Finally, with the help of an evaluation, the section shows that the method is beneficial and motivates to share more experiences. The section concludes by showing that the organizational costs for this method are not very high.

6.2.1. Description

Similarly to the *Observation Sheet* approach (see Section 4.4), the main goal of the Observation Widget with Heuristic Support (short OWHS) is to collect short and ad hoc experiences.

Contrary to paper observation sheets, OWHS is designed as a quick, unobtrusive and low effort means to create *digital* observation sheets. This saves administration effort to convert them into machine readable text, e.g. through OCR software or manual typewriting.

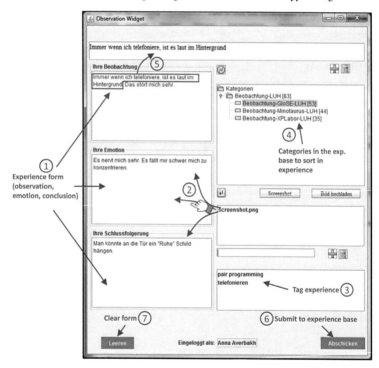

Figure 6.8.: Graphical user interface of OWHS.

The collection method is supported by a desktop client that provides a form to enter a textual description of an experience based on the components of an experience (see Definition 2.4). To lower the cognitive load, the interface is kept simple and textual. A screenshot of the form is displayed in Figure 6.8. OWHS provides several features to ease the writing process and at the

128

same time helps to create a less rawer experience than with the *Observation Sheet* method.

Besides a form for textual description of the experience ①, OWHS includes allows to add screenshots and pictures inside the text ② as well as tags ③ that help to illustrate and specify the experience. Including a picture can often save a lot of descriptive text and writing time. The pictures can be inserted into each of the text fields by drag and drop gestures. This interaction concept should be familiar for the users. To comply with the structure of the experience base (introduced in Section 6.3), the author must select a category to sort in his experience ④. To save the effort to type a caption for the experience, the tool automatically sets a caption ⑤. Therefore, OWHS selects the first sentence of the observation part. I assume that the first sentence of the observation part indicates an event that may be interesting and distinguishing enough for potential readers in the experience base. At the bottom of the form are placed a submit button ⑥ and a possibility to clear the content ⑦ in order to redo the writing. After the upload, the author is presented the URL to the uploaded experience in the experience base for examination and editing (not in the screenshot). To achieve availability and unobtrusiveness like the paper counterpart, OWHS is always one click away. In the current implementation it is placed in the MS Windows taskbar and is always ready to use for the experience bearer.

Heuristic Support for Experience Writing

A central goal of OWHS is to increase the quality of written experiences, i.e. decrease their rawness in comparison to observation sheets. This section presents a concept to decrease experience rawness beginning with characteristics of a *good* experience for an experience engineer and then presenting heuristics to enhance experience quality. The rest of this section is an excerpt from Averbakh et al. [22].

We define *completeness, observability, readability or understandability, traceability* and *verboseness* as characteristics of a *good experience* in the software engineering domain. To avoid misconception, verboseness in this context is meant positively in the sense of being detailed. Our quality aspects are based on established quality characteristics of codified knowledge [74] (accuracy, readability or understandability, accessibility, currency, authority or credibility), requirements [256] (complete, consistent, correct, modifiable, ranked, testable, unambiguous, valid, verifiable) and bug reports [242, 222, 39] (focused, observable, readable, verbose). Requirements and bug reports are similar to an experience in the way that they describe events or make proposals, and are created with the goal to be easy to process in a software development (or maintenance) process. Besides, a bug report can be viewed as a special type of experience or feedback [162]. We did not consider some very bug report-specific artifacts like stack trace, patches, code samples, test cases, etc. [39]. Considering that a goal of our concept is to stay light-weight in terms of effort, we do not want to impose too many restrictions or demands on the experience bearer but also want to ensure the most important aspects of a good experience. Too many demands may discourage experience bearers fearing too much effort to fulfil them. Thus, we discarded the quality aspects *focused*, *unambiguousness*, *consistency*, and *correctness*, because experiences are subjective and impressions of the same event can differ depending on the point of view. We also discarded *verifiability*, *validity*, *ranking*, and *modifiability*, as we consider ranking, rephrasing and comparing experiences with others experiences engineer's tasks. Also *accessibility*, *currency* and *credibility* can be ensured during experience engineering and maintenance and do not need to be imposed on the experience bearer.

Table 6.5.: The heuristic critiques for writing an experience. The column *Helpfulness* displays the percentage of test subjects that found the critique very or rather helpful in the evaluation [22].

ID	Characte-ristic	Criti-cality	Feedback	Heuristic rule	Helpful-ness
1	Complete	Error	The observation / emotion / conclusion does not contain text.	The observation / emotion / conclusion field is empty.	70%
2	Readable, Verbose	Warning	Your observation/ conclusion is very short.	Observation / conclusion text contains < 8 words.	90%
3	Readable	Warning	Your observation/ conclusion could be hard to read.	The readability index is < 30.	40%
4	Observable	Warning	Your emotion might not be clearly stated.	The text in emotion field does not contain an emotion word.	70%
5	Complete, Observable	Warning	Your conclusion might not convey a recommendation.	The text in conclusion field does not contain modal verbs.	60%
6	Observable, Traceable	Warning	Your conclusion possibly lacks actors or persons in charge.	The text in conclusion field is written in passive.	50%
7	Readable	Warning	Your observation / emotion / conclusion contains acronyms.	The text contains at least 2 successive capital letters.	30%
8	Traceable	Hint	The author is missing.	The author field is empty.	30%
9	Traceable, Verbose	Hint	Observation is...	-	60%
10	Traceable, Verbose	Hint	Emotion is...	-	60%
11	Traceable, Verbose	Info	Experience consists of an observation, emotion and conclusion.	-	70%

In Table 6.5 we list the concrete heuristic critiques for experience writing derived from or implementing the characteristics. For each critique the table displays characteristics it supports or enables as well as a criticality. *Error* is the most severe criticality, followed by a *warning*. *Hint* signifies an optional advice and info a general *information*. Feedback can either be general or specific to the experience part currently edited. All critiques, except 11, are context sensitive and are visible within a certain experience part. The last critique being a general information on how to write an experience is visible in all fields. Heuristics 1, 2, 3 and 8 are applied on observation, emotion and conclusion separately but are consolidated in this overview. We implemented the heuristic rules for German language, but they are generally applicable to other languages. To determine the threshold for Heuristic 2, we took a value slightly above the average sentence length of 7.08 words in German literary prose [38]. For the readability measure in Heuristic 3 we implemented the Flesch Reading Ease score [149]. We included this algorithm to raise the experience bearer's awareness about writing style, even though we were expecting low performance because of the brevity of most observations and conclusions. This was successful, as we could observe participants to actually rewrite texts into shorter and less nested sentences during evaluation. For Heuristics 2 and 3 we intentionally did not monitor the emotion field. In our experience, it is not important for an emotion to be very detailed. In Heuristic 4 we use emotion words from the WordNet Affect list[3]. The graphical user interface of the form does not match the one in Figure 6.8. For the evaluation of the heuristic critiques, we developed a separate prototype based on the Heuristic Requirement Assistant (HeRA) [152].

6.2.2. Light-weightedness Classification

According to Definition 4.3 OWHS fulfills C2-C5. The widget is well integrated into the MS Windows environment as it lies in the taskbar as a tray icon, unobtrusive and only one click away (C2). It does not take much time to enter a short observation (C3). The cognitive load is reduced as the bearer can externalize the experience as it happens and does not have to remember it (C4). Besides, the widget is one click away and always available reducing the cognitive load to open a new program and switch context. C5 (shift experience engineering away) is also fulfilled, since this method is part of an experience management process conceived according to the Experience Factory paradigm [23, 30].

Time to share

The user of OWHS only needs to open the tool from the taskbar, write the experience and send it. Adding tags or pictures is not required and thus not considered here. Table 6.6 provides an overview over the components of the time to share, which results in $t_M = \frac{150 \cdot 1}{150} \cdot 7{:}04\text{min} \approx$ 7min. The times for opening and closing the tool can be neglected in this duration range.

In comparison to the Observation Sheet method [230, 213, 148], the experience bearer has to invest ca. 2min more to write one short experience. This is explained in the following:

$t_{i,M}$: We conducted an evaluation of the heuristic critique concept to support experience writing [22]. This evaluation is presented in more detail in the section about benefit. At this

---

[3]We used the words set from http://www.cse.unt.edu/~rada/affectivetext/.

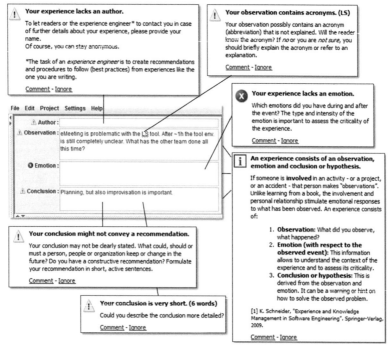

Figure 6.9.: The experience elicitation form with an experience as input and heuristic critiques that would appear beside the form.

point, it is only relevant to note that we observed that writing an experience with heuristic support takes 2min longer than without. The participants spent additional time on reading the critiques and adjusting their experience.

f : I assume that the expected frequency is the same as for Observation Sheet.

e : I also assume that the expected experience size is the same as for Observation Sheet.

p : Like for Observation Sheet, the creator of the experience artifact is a single person.

Cognitive Load

The overall cognitive load of OWHS is a bit higher than for Observation Sheet (see Table 6.7) due to the implementation of OWHS as a standalone tool. The reason against a Web form was

Table 6.6.: Time to share t_M for OWHS.

Method M	Steps and $t_{i,M}$	f	e	p	t_M
OWHS	(i) open widget: 2sec; (ii) write short experience: 7min; (iii) click on "Send": 2sec	150	150	1	7min
Observation Sheet	(i) write short experience: 5min	150	150	1	5min

not to impose the experience bearer to open an additional browser tab. The time to open a browser, find the URL (e.g. as a bookmark) and login takes longer than to open the tool from the taskbar (one click). I also decided against an email client or IDE plugin: Having the entry form as a plugin would mean a contextual connection to the environment and could only be used if the system is open. Experiences that should be written in an OWHS can be more general apart from problems concerning the plugin environment. Thus, the OWHS is decoupled from any specific systems similarly to paper observation sheets. A similar approach that is fully integrated in a SOA environment is described by Lübke et al. [162].

Table 6.7.: Cognitive load for OWHS compared to Observation Sheet.

Method	Recall	Mental demand	Tool integr.	Process integr.	Ad hoc sharing	Cognitive load
OWHS	0	1	1	1	0	3
Observation Sheet	0	1	0	1	0	2

Concepts or rules to recall: The experience bearer does not have to recall any syntax, rules or processes to start operate the tool and to write the experience.

Mental demand of a task: There is some mental demand for the experience bearer. He has to write a short text. The mental demand is slightly higher than for Observation Sheet. This gradation is not reflected in the rough scale, though.

Integration into present tooling: Though it is a standalone tool, it uses well known interaction concepts (like drag and drop for pictures, etc.).

Integration into work processes: OWHS is placed just one click away in the taskbar. I assume that opening and writing a short experience is only a minor change in the usual workflow.

Ability of ad hoc sharing: OWHS can be used anytime assuming that the author is at the workplace.

Benefit

To show that OWHS is perceived as beneficial, we conducted an evaluation. This section is an excerpt from our work [22]. The identifiers of the research questions below are renamed.

Our research goal is to lower the impact of individual communication skills, improve the effort-benefit ratio, and give experience bearers general support on how to submit their experiences. We evaluated four research questions:

1. Do critiques help to write experiences with higher quality for experience engineering?
2. Do users find experiences written with the help of critiques more helpful/richer in content?
3. Do users write more experiences with our critique support than before?
4. Do software engineers find our critiques helpful and justified?

Ten graduate students with a Bachelor of Science degree in computer science and software development experience participated in our evaluation. We specifically chose participants with recent software engineering (SE) experiences, for example students, who had recently participated in SE activities like writing a specification or developing a mobile application in a team. 30% of our participants work or have worked in industrial software projects beside their study.

We conducted a cross validation with two groups (G1 and G2) to evaluate the helpfulness and acceptance of our heuristic support. The participants in G1 were asked to write 2-3 SE-related experiences without and then with our critique system. The control group G2 had a reverse order of experience writing: first with and then without critiques. To answer research question 1 and assess the quality of the written experiences, we specified experience quality metrics. These metrics are very similar to the heuristics. Each experience was read by an experience engineer and evaluated if it complied with the metrics. The experience engineer however, did not know which critiques were fulfilled or not for this experience. Afterwards, in order to evaluate research questions 2, 3 and 4, both groups answered a questionnaire (see Appendix G) about the helpfulness of the critiques in general and for each critique in particular. They also had to rank four experiences according to their perceived quality. Two of the experiences were of poor and average quality according to our metrics. They lacked a conclusion, contained very specific acronyms, or were too short to convey enough information. The other two were the same ones but enhanced with our critique system. We used real experiences from a past distributed SE project.

The boxplot in Figure 6.10 shows a visible experience quality improvement for research question 1. The median rates of the fulfilled quality metrics for experiences created without feedback (G1 and G2) are distinctly lower in comparison to those with critiques (G1c and G2c). The boxplot for G2 indicates a slight learning effect compared to G1. The median of G2 is slightly higher than of G1 and the quality improvement between G2 and G2c is smaller than between G1 and G1c.

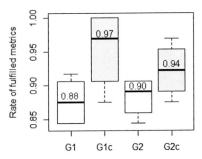

Figure 6.10.: Experience quality without (G1, G2) and with critiques (G1c, G2c). Note that G1 started without, while G2 started with critique support.

Research question 2 can also be answered positively. In Figure 6.11, 80% consider experiences enhanced with the help of critiques more helpful. Experiences enhanced with the help of critiques (E2 and E4) were ranked higher than the original ones (E1 and E3). The poor experience E1 (very short, acronyms, no conclusion) was unanimously ranked as of bad quality.

Research question 3 was confirmed. 70% would write more experiences with a tool that has critiques support. For 50% critiques could lower the barrier and motivate to write more experiences. The latter is a rather inconclusive result. It indicates that users should be able to deactivate the heuristic critiques.

Research question 4 was also confirmed. The statistic for research question 4 shows that 100% of the students found the critiques (rather) helpful and 80% justified. The critiques made all participants feel more certain how and what to write in an experience. On the other hand, the heuristic feedback was perceived as rather interrupting (70%). The participants often had to stop writing to read the feedback text. For 70% of the test participants, feedback slowed down their writing process. We think however, that this negative effect can vanish with repeated use and growing familiarity with the critiques.

For the discussion of possible threats to validity of this experiment refer to Averbakh et al. [22].

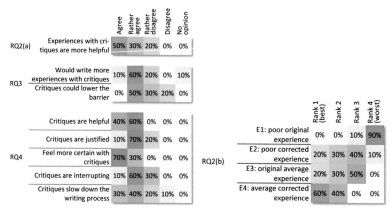

Figure 6.11.: Survey results.

Organizational Cost

The organizational cost for OWHS in comparison to an Observation Sheet is lower (Table 6.8). Since the experiences are digital and are automatically sent to the experience base, no manual uploading or typewriting of the written text is required. OCR algorithms for handwriting recognition do not have a good enough precision (e.g. [145, 52]), thus it would still need a manual check.

Table 6.8.: Organizational cost and experience quality of the OWHS compared to Observation Sheet.

Method	Min. EE effort	Admin. effort	Maturity	Rawness
OWHS	3	1, 2	2	4
Observation Sheet	3	1, 2, 5	2	5

Considering a distributed and collaborative project, the minimal experience engineering effort is equal to the one for Observation Sheet. Experiences must be checked, if they contain sensible information. In case the observation sheets are only scanned and not typed, the effort to analyze may be even higher for Observation Sheet. Machine-readable text can have a programmatic anonymization support. I have employed a tool[4] to anonymize LIDs that were created after each

---

[4]This anonymizer tool was created in one of the software project courses.

136

software project course at the Software Engineering Group. The anonymizer tool recognized project and person names and replaced them with dummy strings. I used these LIDs to create a software project experience base for students[5]. The administration effort is lower than for the Observation Sheet method, since uploading experiences (EE task 5) is not required any more. The maturity of the written experiences is the same as for the Observation Sheet method because it is still a personal experience that was not discussed in a group. The rawness, however, is lower than in an observation sheet due to the heuristic feedback. As the evaluation above shows, experiences created with OWHS have quality characteristics (complete, traceable, readable, better granularity and size through verboseness) that are prerequisite for rawness score 4.

OWHS, compared to Observation Sheet, shows that lowering organizational cost and rawness, while increasing benefit means a rise of personal effort and cognitive load. This is however, a relatively small increase of effort for an experience authoring method in the *light-weight* classification range. Besides, OWHS returns a noticeable (perceived) benefit on personal and organizational level. Moreover, adhering to the heuristic critiques is not obligatory, which would reduce the time to share (but also the experience quality and benefit). I expect a learning effect after longer use. The experience bearer should become acquainted with the critiques and consider them from the start. This effect could be already observed in our evaluation [22], where participants wrote slightly better experiences without heuristic help after they had used the editor with heuristic support (G2 in comparison to G1 in Figure 6.10). Optimally, they will automatically compose experiences of better quality without losing time.

The next section presents the experience management process and the experience base for global software engineering that stores experiences collected with AnnoEx and OWHS.

6.3. Experience Base with Rights Management for Global Software Engineering

Besides collection, every EM initiative should have an engineering, dissemination and experience activation strategy. This section presents a concept of an experience engineering (EE) process and an experience base (further referred to as GloSE Base) that is conceived specifically for distributed development projects. The above presented collection methods are connected with the experience base and upload experiences automatically. This section examines how the collection methods from previous sections interact with the GloSE Base with the appertaining EE process and views.

We defined an EE process that takes into account that experiences can contain sensible in-

---

[5]http://ramus.se.uni-hannover.de/SWPBase (in German)

formation and cannot be directly disseminated to other project partners (sharing barrier 1: *NDA policies and sensitive content*). Figure 6.12 illustrates the EE process. It is very similar to the nine EE steps defined in Section 2.1.3. The description is based on our previous work [23]. First, raw experiences are collected during the project on each partner site independently. The

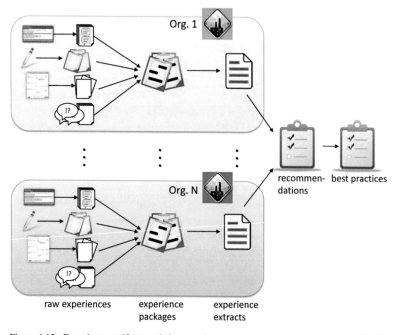

Figure 6.12.: Experience artifacts and the experience engineering process around the GloSE Base (based on [23, Figure 2]).

raw experiences can be annotated documents uploaded with AnnoEx, paper observation sheets or experiences composed with the help of OWHS. Finally, there can be comments to already uploaded experience artifacts or other project-related documents. At each partner site these artifacts are anonymized and rephrased creating *experience packages*. Afterwards, the experience engineers of each partner separately create *experience extracts* from experience packages. These are documents containing consolidated excerpts from recurring or remarkable passages. They are also categorized. At this point the dispersed experience extracts have to be gathered to derive *recommendations* based on experiences from all partners. After applying these recommendations, they can be promoted to *best practices*.

A further goal of the GloSE Base is to prevent information overload and make the base's contents more useful considering different user groups interacting with the base. People react strongly when they are presented an "inappropriate style of information" [212]. We identified 7 different user groups. Without describing each group in detail (see [23]), these groups have different view rights on the various experience artifacts. For example, users who only want to get a rough overview of the base's topic or specifically look for immediate support should not be presented raw experiences but only recommendations and best practices. On the other hand, experience engineers should be able to see all experience artifacts.

At the same time we implemented a rights management system in the base to allow for storing sensible raw experiences, experience packages and experience extracts. A detailed presentation of the access for each artifact type and user roles can be found in our previous work [23].

The GloSE Base with rights management and the EE process were employed and evaluated in the context of a joint research association on global software engineering [28]. We collected experiences about challenges in global software engineering projects. The base has grown since the first evaluation [23]. At the moment, GloSE Base contains 188 raw experiences, 49 recommendations and 1 best practice. Technically, GloSE Base is implemented as a Semantic MediaWiki[6] with Halo Access Control List plugin[7] for rights management.

Altogether, with the above introduced experience collection methods, this tooling environment and processes present a possible EM initiative. At this point it must be kept in mind that a successful EM endeavor must support all steps of the experience life cycle (Figure 2.1). Here I only described the collection and engineering concepts, while active dissemination and activation were not in the main focus of this research. As a quick solution we used newsletters to disseminate new experiences to project partners. We assumed that activation of new experience happens automatically through application of the existing ones.

The next section shows which of the experience and knowledge sharing barriers can be overcome by the AnnoEx, OWHS and GloSE Base concepts.

6.4. Lowered Experience and Knowledge Sharing Barriers

The EM initiative covers all main barriers and some sub-barriers. Figure 6.13 depicts these barriers. The circles next to the barriers represent the solutions from the EM initiative described in this chapter. I assume a barrier category as overcome, if at least one of its sub-barriers is

---

[6]http://semanticweb.org/wiki/Semantic_MediaWiki
[7]http://www.mediawiki.org/wiki/Extension:Halo_Access_Control_List

overcome. Otherwise, I describe the mitigation of the barrier category itself.

An EM system consisting of GloSE Base, AnnoEx and OWHS including the EE process does not overcome all sub-barriers from the complete cause effect overview in Figure 3.1, though. The barriers here are a superset of the barriers overcome by light-weight methods in general (Figures 4.2 and 5.8). Next, follows a description how each barrier from Figure 6.13 was overcome in detail. Sections 6.4.1 - 6.4.6 stand for each barrier category. The barriers with italic font are sub-causes of this barrier category. The IDs of each barrier refer to the IDs in the barrier list in Appendix A.

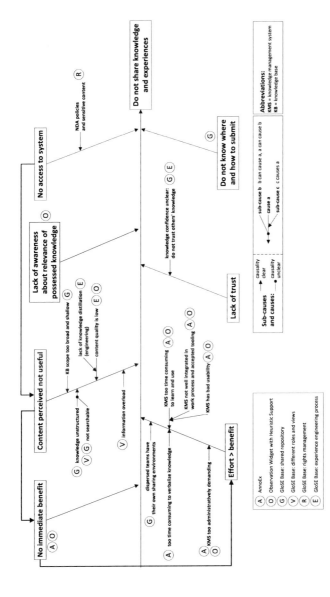

Figure 6.13.: The sharing barriers that are mitigated by AnnoEx, OWHS and the GloSE Base with rights management and EE process.

6.4.1. No Immediate Benefit (ID:17)

According to the evaluations in Sections 6.1 and 6.2, AnnoEx A, AnnoEx B and OWHS return an immediate benefit. Sharing and presenting annotations was considered useful and the heuristic support in OWHS was perceived as motivating to share more experiences.

6.4.2. Content Perceived Not Useful (ID:18)

Knowledge Unstructured (ID:59): In the GloSE Base, the experiences are structured according to their rawness. Recommendations are structured according to the use cases in a distributed development project, e.g. how to communicate, how to plan or how to develop.

Not searchable (ID:60): Being a MediaWiki, GloSE Base contains a default search functionality.

KB scope too broad and shallow (ID:20): The scope of the GloSE Base is quite specific. We collected experiences on how to conduct globally distributed project.

Lack of knowledge distillation (engineering) (ID:24): The above presented experience engineering and management process overcomes this barrier.

Content quality is low (ID:19): The experience engineering process and the creation of recommendations that can be promoted to best practices aim at augmenting content quality. Besides, creating recommendations from joint experiences can make them more valuable with new insights than having separate EM initiatives. But even on the level of raw experiences, the evaluation of OWHS has shown that heuristic critiques can raise the quality of written raw experiences.

Information overload (ID:62): The different views (and view rights) and roles should prevent that irrelevant information is presented to users with different goals. The presentation structure of recommendations according to use cases should make the base more useful.

Lack of awareness about relevance of possessed knowledge (ID:24): OWHS aims at overcoming this barrier through heuristic support. Stimulating reflection-in-action, feedback can raise awareness about one's knowledge and relevance of specific facets.

6.4.3. No Access To System (ID:16)

NDA policies and sensitive content (ID:1): The rights management concept of the GloSE Base prevents that sensible content is accessed by other parties. At the same time, anonymized and rephrased experiences can be shared with partners. This should prevent frustration about a lack of access and raise the benefit.

6.4.4. Do Not Know Where and How to Submit (ID:11)

Having a shared base overcomes the lack of awareness where and how to submit experiences [23]. Of course, its existence must be sufficiently advertised.

6.4.5. Lack of Trust (ID:9)

Knowledge confidence unclear: do not trust others' knowledge (ID:8): The recommendations and best practices in the GloSE Base have links to raw experiences they were derived from as prove and source. This should raise the confidence of a recommendation.

6.4.6. Effort > Benefit (ID:52)

Knowledge management system (KMS) too administratively demanding (ID:54): Both light-weight collection methods are easy to operate and install. GloSE Base is set up by an experience engineer.

Too time consuming to verbalize knowledge (ID:47): AnnoEx is a knowledge and experience collection method that does not require additional knowledge verbalization.

Dispersed teams have their own sharing environments (ID:15): A shared experience base and a uniform experience engineering process, if accepted, can overcome the problem of isolated EM solutions.

KMS too time consuming to learn and use (ID:56): Both collection methods (and light-weight methods in general) adapt well to the work processes and tooling environment. This should accelerate the learning process.

KMS not well integrated in work process and accepted tooling (ID:50): Both collection methods are well adapted to the present tooling and processes.

KMS has bad usability (ID:53): AnnoEx and OWHS were designed to support and facilitate the most important use cases and user goals.

6.5. Contributions

This chapter evaluates **RQ4:** *Do light-weight experience collection methods lower the significant experience and knowledge sharing barriers and increase participation in an experience management initiative in a distributed software engineering project?*

The evaluation of the very light-weight (as classified by the measurement system constructed in this thesis) text analysis method AnnoEx confirmed that light-weight approaches also benefit project participants by saving them time to search for specific knowledge and experience in document annotations.

The second case study of the light-weight experience authoring method OWHS was perceived as helpful as well. The study participants also considered the presented approach of sharing ad hoc experiences with immediate feedback based on heuristic critiques as motivating to write more experiences.

Together with the experience engineering process and the experience base with rights management, both experience collection methods can overcome all main knowledge sharing barriers.

7. Related Work

This section presents related research that covers one or more aspects of the concepts in this thesis: the notion of light-weightedness, measurement and assessment of experience collection methods, incentives to motivate experience sharing and other literature reviews on sharing barriers.

7.1. Characterization of Light-weight Experience Collection

This section presents related research that covers one or more aspects of light-weight experience collection as characterized in this thesis.

The *Experience Factory* concept defines a set of processes to support experience reuse [30]. Basili et al. postulate that knowledge and experience have to be packaged and analyzed for reuse potential. They propose to employ an organizational unit that solely manages experience. This unit is responsible for capturing experiences, storing and packaging them. Packaged experiences should flow into project planning activities. The suggestion to introduce a dedicated unit for experience engineering resembles criterion C5 of light-weight experience collection (Definition 4.3 on page 35): defer EE tasks to others than the experience bearer. In contrast to this definition though, Basili et al. present an organizational structure and do not offer solutions to lower the effort for experience bearers on a personal level in a distributed software development project.

Basili and colleagues also define an experience management approach, called *Dust to Pearls* [31, 33]. They adapt the *Experience Factory* approach to serve their model. *Dust to Pearls* suggests that a successful EM initiative must deliver short-term benefits (dust) and evolve to return long-term benefits (pearls) to experience bearers and organization. Dust represents the experiences that are easily captured and quickly disseminated to ensure immediate benefit to those, who share their experience. Pearls are results of longer-term and high effort experience engineering. Realizing that proper experience engineering activities, as recommended by the *Experience Factory*, are very time-consuming, *Dust to Pearls* focuses on returning an immediate benefit to experience bearers as well. Basili et al. mention two approaches that return *dust* [32, 33]: *eWork-shop* and *Answer Garden*. Both methods are classified as (borderline) light-weight in this thesis. However, the *Dust to Pearls* approach itself only explicitly describes the necessity of an immediate benefit as the most important motivator for experience sharing. Immediate benefit, though returned by all identified light-weight methods in this thesis, is not a necessary prerequisite according to Definitions 4.1 and 4.3 (pp. 34, 35). The notion of light-weight experience collection presented here focuses on the other end of the effort / benefit equation: effort reduction.

Dingsøyr et al. compare two post-mortem experience elicitation methods (*Experience Reports* and *Postmortem Reviews*[1]). They mention the necessity to employ efficient experience collection methods "that do not require a lot of effort" [87]. This understanding is similar to the light-weight notion in this thesis. However, they do not present any systematic definition or measure how to determine light-weightedness.

Grudin formulates eight fundamental challenges for successful adoption of tools in the field of Computer Supported Cooperative Work [110]. Some of Grudin's challenges are explicitly and implicitly approached in this thesis. Among other things, he mentions the problem of work and benefit disparity and the need for unobtrusive accessibility of the tool support. This thesis transfers these challenges to the experience management field and presents approaches that have the goal to lower the effort for learners and providers. The unobtrusive accessibility is realized in e.g. *Observation Widget with Heuristic Support* (see Section 6.2).

Stenmark and Lindgren [235] suggest design principles for KM systems that are similar to the five light-weight characteristics in this thesis (Definition 4.3 on page 35). They apply Grudin's design challenges [110] to knowledge management. Specifically, they state that many EM initiatives operate mainly to benefit the collective and not the individual, which can lead to a fast decline of interest towards the experience management system. Similarly to this thesis, Stenmark and Lindgren point out consequences for each of Grudin's challenges and explain them. However, they only give short and abstract advices and do not provide specific solutions to the effort / benefit dilemma.

Schneider [212] defines light-weight experience similar to this thesis. He also takes Grudin's challenge about unbalanced effort and benefit [108] as basis to motivate light-weight approaches. He points out that it is necessary to consider the *perceived* cost and benefit on the personal level, i.e. the experience bearer. Schneider describes a light-weight approach as one that may return a "limited benefit" but also demands low personal effort. In his book [216, p.170] he provides a definition of light-weight and heavy-weight experience stating that heavy-weight methods aim to return much benefit but at the same time demand much effort, while light-weight methods limit effort and benefit. The definitions of light-weight and heavy-weight collection methods in this thesis (Definitions 4.1 and 4.2) are based on his definition. I include his statements but separate both notions. While Schneider's definition [216, p.170] does not differentiate personal

---

[1]This method is abbreviated as LPMR in this thesis and is classified as borderline light-weight. Refer to Section 4.4.6 for its description.

and organizational effort, I demand of a light-weight experience collection method to explicitly lower *personal* effort. To operationalize this notion, I provide criteria (Definition 4.3) and concrete measures for qualitative and quantitative assessment of light-weightedness.

Maalej and Happel propose five enablers of light-weight knowledge sharing support in distributed development teams [163]:

Contextualization: The sharing environment should be able to identify context similarities and map them. The knowledge context should be traceable to determine where the knowledge arose from.

Personalization: The search system of the sharing environment should return personalized results in order not to overload the user and not to decrease its efficiency.

Proactive Assistance: The knowledge management environment should proactively identify situations where knowledge is needed and recommend suitable knowledge artifacts, e.g. of a similar context.

Decentralization: Sharing systems in a distributed setting should be decentralized to ensure scalability, heterogeneity and flexibility.

Integration: Integration in the work environment should reduce context switches.

Unlike this thesis, these attributes do not specifically concern experience collection. They are general success characteristics of a knowledge management system in a distributed environment. The aspect of integration is reflected in the definition of light-weightedness criteria in this thesis. The other aspects do not refer to experience *collection*. Contextualization is a requirement for experience quality. Personalization and proactive assistance are issues related to the dissemination step. Decentralization is a rule to be applied on the general design of the system, which is outside of the thesis scope. Furthermore, Maalej and Happel employing the notion, neither define what *light-weight* is, nor mention personal effort reduction for experience collection as a goal.

7.2. Measurement of Knowledge and Knowledge Collection Methods

This section presents approaches to measure experience collection methods (or broader: EM initiatives) and effort.

As a measurement approach, Schneider formulates an equation to measure the utility of an EM initiative [212]: *Utility = Benefit / Effort*. Utility is the comparison of benefit that the EM initiative returns and the effort put into the initiative. Thereby, he notes that it is important to consider the *perceived* effort and benefit as it can vary depending on the situation, task and person. Schneider draws implications from this equation, which are close to the light-weight definition. These implications were discussed in the previous section of this chapter. Schneider does not go into more details of this formula. He does not provide concrete values or applies the equation to the methods he introduces in the publication. The equation merely serves to illustrate and motivate light-weight EM solutions. Additionally, this formula calculates the general utility of an EM. It does not specifically make a statement about *light-weightedness*. The metrics in this thesis, however, specifically reflect if and to what extent an experience collection method is light-weight.

Cooke et al. conducted a review on metrics to assess team knowledge [66]. They discuss measurements of similarity, i.e. consensus or overlap, of knowledge within the team. This can be measured by counting similar answers, for example. Further they mention measurements of accuracy of team knowledge, which are the number of correct responses and information coverage within the team (scaling individual weights). Unlike the focus of this thesis, these metrics should predict team performance and not light-weightedness of a collection method.

To measure cognitive load, there are approaches from the psychological domain. Marshal measures cognitive activity by observing pupil dilation during a task [168]. This method to estimate cognitive load is not practical, though. It is very invasive and cannot be conducted outside a lab.

7.3. Catalogue of Light-weight Experience Collection Methods

Many researchers have analyzed and compared knowledge and experience collection methods. Some researchers attempted to set up taxonomies about collection methods. Hoffmann et al. present a taxonomy categorizing elicitation methods into three categories: analysis of tasks that the expert performs, various types of interview, and contrived techniques [125]. Gavrilova and Andreeva also compare various elicitation methods [102]. Their focus lies on presenting a taxonomy and comparing their suitability to elicit a different kinds of knowledge: tacit, explicit, individual or group knowledge. Unlike this thesis, the above presented taxonomies only limit their focus to knowledge elicitation methods, i.e. interviews, questionnaires and workshops. Furthermore, they do not consider them in terms of light-weightedness.

Cooke presents a detailed analysis of various knowledge collection methods [67]. He sorts them according to their method family: knowledge elicitation methods, process tracing, conceptual methods. For each method family, he presents a hierarchical list of collection methods. He compares the methods according to formality, obtained data, timing of elicitation, directness of expert response and elicitor role. Cooke presents a different set of collection methods and does not examine the light-weight criteria.

Burge elaborates on an exhaustive list of knowledge collection methods that also cover other methods than elicitation, e.g. document analysis and prototyping [46]. Like the works above, she does not examine them under the same aspects as does this thesis.

Several studies compared elicitation methods in terms that are also examined in this thesis. These terms are *time spent on sharing* and *time for preparation and transcription* [47, 35, 184, 159, 48]. These aspects resemble *time to share* (see Section 5.3.1) and *experience engineering effort* (see Section 5.4.1). However, these studies only examine elicitation methods. Moreover, some of these methods were not applied to elicit knowledge or experience, but marketing [35], requirements engineering [184] and information systems [159]. Those studies that are in the knowledge management field [47, 48] rather focus on determining if the methods were efficient for the knowledge engineer. As opposed to concepts of this thesis, Burton et al. suggest that it is "worth bearing a little discomfort" [48] in order to make the elicitation and steps more efficient afterwards.

Summing up, the related work on analyses of knowledge collection methods do not examine all criteria of light-weight experience collection and do not examine the same set of collection methods.

7.4. Incentives for Experience Sharing

Beside light-weight approaches to lower sharing effort, there are strategies to raise participation in an EM initiative through increase in motivation.

The *Self-determination Theory* describes different types of motivation: intrinsic motivation, extrinsic motivation and amotivation [207]. People are intrinsically motivated to perform an activity, if they do it for the activity's sake only. They can also be extrinsically motivated to do it. Extrinsic motivation has four gradations that range from almost self-determined to more and more externally regulated with less self-determination involved. The more the activities are motivated toward intrinsic rather than extrinsic direction, the more enjoyable and satisfactory they are to the person. The third motivation type is amotivation. It happens if a person has no interest in performing the task and there is no regulatory apparatus to motivate him. Many techniques have been researched to increase motivation.

One way to externally motivate knowledge sharing is through reward mechanisms. This is a large field of study. For example, Bartol and Srivastava discuss organizational reward mechanisms [29]. They suggest that sharing knowledge should be rewarded and that such mechanisms are relatively easy to accomplish, since knowledge value can be measured. Dencheva et al. propose rewards in form of reputation to raise the intrinsic motivation to share knowledge in a

Wiki-based repository [80]. They use ranks, awards and levels to achieve higher participation and higher quality of articles. Similarly, Gagné developed a model of knowledge-sharing motivation [100]. She presents an number of hypotheses on how motivation can be increased. She strives to raise motivation, through increasing interest and joy [100]. Singer presents methods to promote adoption of software engineering principles and patterns for non-developers [228]. In his dissertation he also discusses methods to raise motivation like points & levels, leaderboard, rankings, appreciation and triggers. Cabrera also discusses interventions and motivational methods to support knowledge sharing [49]. Pointing out that non-monetary rewards, such as e.g. social recognition, can be powerful motivators, he argues that rewards can also be "gamed". People may artificially inflate the number of contributions and disregard quality for the sake of receiving the reward.

The concept of *Serious Games* can also serve the purpose to increase motivation for tedious and demanding tasks (e.g. [9, 180]). A serious game is defined as a game that serves other purposes than solely to entertain [238]. It can be a (computer) game or playful process elements that address serious topics often with educational purposes (e.g. in healthcare, military, mathematics, software engineering practices). The above introduced social mechanisms (ranks, awards, etc.) can be incorporated into the work process to make it more game-like. For example, Singer and Schneider employed gamification mechanisms to motivate a software engineering practice (increase number of code commits to a version control system) [227].

The above presented incentives have an opposed approach considering the effort / benefit relation of a method (also called *utility* according to Schneider [212]) than the approach of light-weight experience collection. Figure 7.1 illustrates this comparison of light-weight approaches and motivational incentives. The concepts in this thesis tries to lower the effort, while motivational incentives try to increase personal benefit from using or contributing to the knowledge management system. Both approaches, however, seek to achieve a situation where the reward from sharing or the base's content exceeds its cost. This is the point where knowledge sharing will occur [147].

7.5. Literature Reviews on Knowledge Sharing Barriers

Several authors have conducted reviews about obstacles to sharing knowledge [196, 27, 88, 176]. I considered these reviews as input in my own review in Chapter 3, since some of them presented issues that I did not find in my review. Examples are barriers 44 and 45 in Appendix A: *difference in experience* and *age differences* [196]. Not including them would make the review less complete.

All these reviews except for Riege [196] did not conduct a systematic literature review. Al-

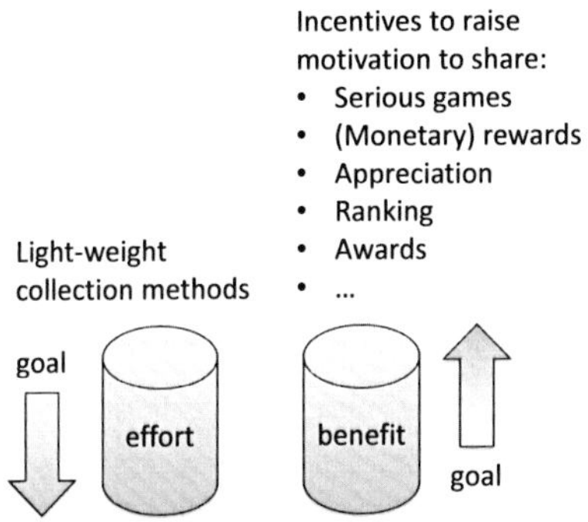

Figure 7.1.: Light-weight collection methods and motivational incentives to share experiences and their focus in the effort / benefit relation of an EM initiative.

together, all of these reviews present a smaller number of issues as the review in this thesis. Moreover, none of the works presents a causal or associative relation between (a subset of) sharing barriers that are relevant to distributed software engineering.

8. Conclusions and Future Work

This chapter first critically discusses the limitations of the concepts in this thesis and the research methods used to create it. The second section provides ideas for future research, followed by a summary of the thesis' contributions.

8.1. Critical Analysis and Limitations

It is important to highlight some of the threats to validity of the work reported in this dissertation.

A limitation of this work is the potential concern for the validity of the empirical evaluation (Chapter 6) presented in this thesis. Due to the low number of only two case studies, a generalizable conclusion may not be drawn. Future work will need to conduct a long-term study in a distributed industrial development project to see if the presented framework for creating a light-weight EM system is successful. However, a qualitative analysis in the case studies provides indications that light-weight experience collection methods are beneficial and can diminish the sharing barriers. Together with the presented experience base including rights and views management and the experience engineering process, I have shown that every major sharing barrier can be overcome.

Further, the evaluation may pose a threat to conclusion validity. The goal of the evaluation was to investigate if light-weight experience collection methods and an experience management system (constructed according to the framework) were perceived as motivating and whether it overcomes the main sharing barriers. However, the setup of the evaluation cannot exclude the possibility that the participants perceived the methods as helpful because of their specific functionality, i.e. to extract annotations or to support experience bearers with heuristic critiques. This perception may be not solely due to the trait of being light-weight.

Futhermore, the evaluation does not confirm the correctness of the five light- or heavy-weightedness ranges in Figure 5.5, nor does it examine possible differences in effects of the methods within these ranges. I only evaluated if a collection method classified as at least borderline light-weight or lower mitigates the sharing barriers as described in Section 5.7. Additional evaluations and comparisons are needed to evaluate the ranges in this portfolio. Further studies should also examine, if the assumed relation of the time to share and cognitive load in the portfolio (Figure

5.5) are confirmed. The gradient of the limits is based on these assumptions.

Even though a survey was distributed to software engineers in the industry (see benefit evaluation of the Annotation Extractor, Section 6.1), the majority of results were gathered by evaluation with computer science students and software engineering PhD students. This may pose a threat to external validity. However, the overall tendency of the evaluations affirms **RQ4**, indicating that experience collection methods lower the significant sharing barriers and increase participation in a distributed software engineering project.

8.2. Outlook

Future work could examine different relationships from Table 5.21 that were not determined in this thesis. Such relationships can be found in experience engineering and administration efforts. It is also unclear how administration effort would change in a situation where a new tool support has to be developed and not reused. Another unclear relationship is the personal benefit measured through time saved (Equation 5.4). The influence of the maturity and rawness factors must yet be determined, e.g. through qualitative studies. Defining this formula would make it possible to measure the utility (see Section 7.2 or [212], where utility is mentioned) of an experience collection method and the EM initiative. Utility could be an instrument to detect effort / benefit disparity in an EM initiative.

Another topic for research could be the causal relationships of the knowledge sharing barriers in Figure 3.1. The causal and associative relationships presented in this thesis are only a hypothesis and are not empirically proven. This could be achieved through a qualitative analysis.

The light-weight experience collection methods introduced in Section 4.4 are presented in their original implementation and organizational setting, which was not in every case a distributed software engineering project. Also the experiments in the evaluation were not conducted in a distributed setting. Future research should focus on experiments in a distributed project context.

This thesis has focused on the question how to minimize effort to share experience. Related work has shown that there are also other approaches focus on maximizing the personal perceived benefit from the EM initiative. This is achieved through social interventions, for example, without much additional experience engineering effort. As ongoing research has proven, social incentives can motivate software engineers to adopt certain processes and tasks (e.g. [80, 228]). Future work could research methods to integrate social reward systems into light-weight experience collection strategies to boost the perceived benefit for those, who contribute their experience.

Lack of trust, besides effort and benefit disparity, was identified as one of the most severe experience and knowledge sharing impediments in distributed projects. The concept of light-weight experience collection does not explicitly provide means to enhance trust, though the experience base and engineering process described in the evaluation (Section 6.3) provide a possible solution to trust related problems. At this point, social or other interventions could be used to increase trust among collaborative teams. These interventions should increase the frequency of communication, raise awareness and make collaboration partners more accessible to each other [127, 186, 131]. However, these methods will not take full effect without a proper management process to build up trust [106, 132]. A healthy and open knowledge sharing culture is important to create "an atmosphere in which organization members feel safe sharing their knowledge" [135].

To better support distributed teams, future research could focus more on the characteristics and distinctions of distributed teams and adapt light-weight measurements accordingly. Matthews et al. [169] point out three types of collaborative teams (without claiming completeness): 1) communities of practice, 2) large, dynamic project teams and 3) task teams. These teams can be differentiated by a list of criteria like collaboration duration, reason for collaboration, team size, etc.. The different characteristics of the groups may have an influence on adoption of experience collection practices and perception of its usefulness. A community of practice, for instance, is a group of people with similar interests and job function [251]. They usually come together to share their knowledge and experience. A priori, these people may have a more positive attitude and motivation towards sharing experience and would invest more effort to do it. On the other hand, members of a project team may have higher barrier to share, as a team usually has a higher member fluctuation and less common interest to work together. For this type of collaboration group, very light-weight (e.g. by-product or data mining) solutions could be more reasonable. A task team is similar to a project team but self-managed, short-lived and more ad hoc in nature. Here, by-product methods like *FOCUS* (see Section 4.4) could be employed. At this point, future work could determine which collection method category is more suitable for which collaboration type. Matthews et al. also propose several collaboration tools that should be employed for the different collaboration types. A possible research direction could be to integrate light-weight experience collection methods into these tools.

8.3. Contributions

Motivated by the problem that experience bearers in distributed software development do not share experiences due to excess of effort for sharing, this thesis contributes an approach to lower personal effort for experience sharing. This is achieved by a light-weight experience collection approach. To this end, research questions in this thesis have focused on defining what light-weight experience collection *is* and how light-weightedness can be *measured*. In particular, this thesis makes four contributions presented in the next sub-sections.

8.3.1. Definition of the Problem Space: Experience and Knowledge Sharing Barriers

The problem space of this thesis is the higher reluctance of software engineers to share experience in distributed software engineering. To investigate state-of-research on experience sharing barriers and confirm our problems setting up a knowledge management initiative in the distributed collaboration project *e performance* [179], I conducted a literature review. As a result 63 barriers could be identified. The subset of these barriers that were symptomatic to distributed software engineering were placed in causal or associative (where causality was unclear) relationships to each other. The main barriers identified in the review are *no access to system, lack of awareness about relevance of possessed knowledge* or *where and how to submit, lack of trust, not useful content, lack of immediate benefit,* and at last, *effort > benefit*. As an central conclusion of this review, it indicates that a lack of time and of balance between effort and benefit are central barriers. This conclusion motivates the need to focus on effort reduction for experience bearers during experience collection.

8.3.2. Framework of Light-weight Experience Collection

In order to overcome the imbalance of effort and perceived benefit to share experiences in a distributed software engineering project, this thesis proposes a framework to create and identify light-weight experience collection methods. Light-weight experience collection methods focus on effort reduction for experience bearers. In this thesis, I first propose a qualitative way to assess light-weightedness through criteria and then provide a measurement system for a quantitative analysis.

Identification Criteria of Light-weight Experience Collection

To overcome the problem of too much effort, I propose to employ light-weight experience collection methods. Though light-weight methods in the experience and knowledge management domain are known and researched, no common and unanimous characterization exist. I have derived five criteria for the notion of light-weightedness for an experience collection method. The main goal of a light-weight experience collection method is to lower personal effort of the experience bearer.

Measurement System for Light-weightedness and Light-weightedness Scale

To allow a more detailed assessment, I define a measurement system of light-weightedness of an experience collection method and provide a differentiated classification scale for light-weight and heavy-weight methods. As basis for the measurement I use the criteria that describe a light-weight collection method. This thesis provides metrics to measure personal and organizational effort as well as personal benefit. Effort and benefit are essentially subjective aspects, as they involve people and their capabilities, skills and opinions. In order to objectively measure them, I provide default values for the most central activities (reading and writing time, etc.). Applying these measures and using the default values, a knowledge manager can assess and predict if and to which extent the collection method he is about to introduce will suffice the light-weightedness definition. These metrics can also be viewed as a guideline to construct a light-weight experience collection strategy.

8.3.3. Analysis and Catalogue of Light-weight Experience Collection Methods

Besides a system of light-weightedness metrics, this thesis also contributes a catalogue of 14 light-weight experience collection methods as a result of a literature review. These methods were assessed according to the measurement system presented in this thesis. This catalogue can be used by a knowledge manager as basis to select an appropriate collection method. The identified light-weight collection methods were analyzed and displayed in a portfolio of light-weightedness. They were used as basis to propose a scaling for different light-weightedness and heavy-weightedness gradations. The assessment also showed that light-weight methods strive to return an immediate benefit, while heavy-weight methods focus on gaining benefit on a long-term. Light-weight methods also do not require exceedingly much experience engineering and organizational effort to produce immediate benefit on the personal level. On the other hand, they usually collect experience of a lower maturity and higher rawness than heavy-weight methods.

8.3.4. Evaluation of the Framework

I empirically evaluated the light-weight measurement system and classification scheme by conceiving two light-weight collection methods for an EM system, including an experience engineering process. The overall system particularly supports global software engineering projects and overcomes all main sharing barriers in a distributed setting. The experiments showed that both light-weight collection methods were perceived useful and returned an immediate benefit to the participants. The results also indicate a raise of motivation to share more experiences.

The contributions of this thesis can also be relevant for co-located development. Evidence indicates that team members in a software development project do not have to be distributed in order to show the symptoms of a distributed project and the resulting challenges to experience management. Already a distance of 30 meters is enough to diminish communication and awareness similarly to a globally distributed project [119, 16].

List of Definitions and Examples

A. Experience and Knowledge Sharing Barriers

Table A.1.: Experience and knowledge sharing barriers.

ID	Category	Barrier	Sources
1	DSE	NDA policies and sensitive knowledge (confidentiality)	[195, 23, 179, 243, 27, 18, 101, 143, 138, 137, 188, 15]
2	DSE	competitive relationship of partners	[179, 196, 27, 92, 237]
3	DSE	different knowledge sharing cultures	[78, 173, 27, 239, 129, 197, 105, 237, 250, 188, 209]
4	DSE	impaired communication (low communication bandwidth and low context)	[115, 120, 196, 129, 190, 237, 188, 209]
5	DSE	lack of awareness about existing knowledge	[23, 163, 196, 143]
6	DSE, org.	knowledge hostile environment	[132, 261, 196, 13]
7	DSE, org.	system is external and used as control mechanism	[179, 60]
8	DSE, org.	Fear / uncertainty about access to knowledge base after project ends	[179]
9	DSE, social	lack of trust	[132, 248, 196, 27, 26, 54, 55, 60, 127, 237, 64]
10	DSE, social	do not trust others' knowledge	[14]
11	DSE, social	do not know where and how to submit knowledge	[217]
12	DSE, social	poor verbal / written communication skills	[196, 237]
13	DSE, social	different national culture and ethnic background	[196]

14	DSE, social	shared language not mastered well enough	[88, 203]
15	DSE, tech.	dispersed teams have their own sharing environments	[179, 193, 27, 250]
16	DSE, tech., org.	no access to system	[179, 105]
17	process-rel.	no immediate benefit	[179, 63, 75, 248, 49, 33, 25, 214]
18	process-rel.	content (perceived) not useful	[179, 27, 25, 60, 218, 142, 45]
19	process-rel.	content quality is low	[25]
20	process-rel.	KB scope too broad and shallow	[75, 216, 218, 220, 179, 214]
21	process-rel.	knowledge becomes outdated too quickly (low currency)	[118, 60, 15, 142], [179] (Wiki)
22	process-rel.	too little seed	[216, 218, 214], [179] (Wiki)
23	process-rel.	experiences are ambiguous: solutions can work in a situation and not work in another	[195, 79]
24	process-rel.	lack of knowledge distillation (experience engineering)	[60, 220, 214]
25	process-rel., org.	communication flows are restricted into certain directions (top-down)	[196, 45, 126]
26	org.	not enough management imposition / commitment / interest	[49, 196, 60, 259, 237, 214, 179, 13, 117]
27	org.	physical work environment restrict effective sharing practices	[196]
28	org.	competitive internal work environment	[196, 49]
29	org.	business units too big and knowledge sharing unmanageable	[196, 15]
30	org.	EM initiative costs (too much) money that could be invested in tasks with clearer returns	[49, 196, 27, 60, 15, 13]
31	org.	no proper rollout strategy, lack of pilot phase	[60]

32	org., social	mistakes are tabooed and not recorded	[132, 196, 63]
33	org., social	lack of social network	[196, 173, 25]
34	org., tech.	strong hierarchy, position-based status, centralized	[196, 88]
35	org., tech.	not enough training and user support	[196, 259]
36	social	not enough time	[179, 80, 49, 196, 25, 138, 191, 250, 188, 142, 241, 124, 117, 209, 206]
37	social	protection of competitive advantages / power, fear to become replaceable	[132, 75, 248, 80, 234, 196, 27, 54, 84, 60, 235, 88, 237]
38	social	resistance to be known as expert	[82]
39	social	fear of knowledge parasites ("free-ride")	[132, 80, 196, 27]
40	social	no reward / acknowledgement	[132, 248, 196, 173, 27, 45, 218, 88, 45, 50]
41	social	evaluation apprehension (fear of misinterpretation or criticism of one's knowledge)	[132, 248, 18, 19, 25, 88, 176, 237, 42]
42	social	lack of perceived self-efficacy (awareness about relevance of possessed knowledge)	[112, 248, 80, 49, 196, 18, 88, 176, 140, 237]
43	social	"not-invented-here" syndrome (create new knowledge more prestigious than reuse)	[141, 132, 216, 237]
44	social	difference in experience	[196]
45	social	age differences	[196]
46	social	differences of education levels	[196, 239]
47	social	knowledge is too tacit to be captured and categorized (too time consuming to verbalize knowledge)	[82, 209]
48	social	knowledge sharing is "bragging"	[25, 157]
49	social, tech.	system is too complex	[179]
50	social, tech.	does not fit into workflow	[179, 75, 49, 196, 27, 218, 220, 235, 259, 105, 203, 13]

51	social, tech.	too much effort	[118, 248, 195, 110, 60, 196, 63, 235, 88, 50, 215]
52	social, tech.	more perceived effort than benefit	[108, 212, 215, 248, 49, 110, 55, 235, 105]
53	tech.	KMS has bad usability / user friendliness	[75, 60, 220, 105, 209]
54	tech.	KMS too administratively demanding	[118, 60, 218, 203, 15]
55	tech.	KMS too technology based	[118, 60]
56	tech.	KMS too time consuming to learn and use	[118, 80, 49, 196, 235, 105, 209]
57	tech.	instability, system not reliable, not mature	[75, 27]
58	tech.	system too slow	[179, 75, 142]
59	tech.	knowledge unstructured	[63, 60, 15, 142]
60	tech.	not searchable	[63]
61	tech.	no context in knowledge / documents	[63]
62	tech.	low bandwidth to KMS	[179, 60]
63	tech., social	information overflow / overload	[216, 163, 15, 142, 209]

Table A.2.: Experience and knowledge sharing barriers sorted according to the frequency (number) of references in the literature.

ID	Category	Barrier	References in Literature
36	social	not enough time	15
37	social	protection of competitive advantages / power, fear to become replaceable	13
1	DSE	NDA policies and sensitive knowledge (confidentiality)	12
50	social, tech.	does not fit into workflow	12
3	DSE	different knowledge sharing cultures	11
9	DSE, social	lack of trust	11
40	social	no reward / acknowledgement	10

42	social	lack of perceived self-efficacy (awareness about relevance of possessed knowledge)	10
51	social, tech.	too much effort	10
26	org.	not enough management imposition / commitment / interest	9
41	social	evaluation apprehension (fear of misinterpretation or criticism of one's knowledge)	9
52	social, tech.	more perceived effort than benefit	9
17	process-rel.	no immediate benefit	8
4	DSE	impaired communication (low communication bandwidth and low context)	8
18	process-rel.	content (perceived) not useful	7
56	tech.	KMS too time consuming to learn and use	7
20	process-rel.	KB scope too broad and shallow	6
30	org.	EM initiative costs (too much) money that could be invested in tasks with clearer returns	6
21	process-rel.	knowledge becomes outdated too quickly (low currency)	5
53	tech.	KMS has bad usability / user friendliness	5
54	tech.	KMS too administratively demanding	5
2	DSE	competitive relationship of partners	5
63	tech., social	information overflow / overload	5
5	DSE	lack of awareness about existing knowledge	4
6	DSE, org.	knowledge hostile environment	4
22	process-rel.	too little seed	4
43	social	"not-invented-here" syndrome (create new knowledge more prestigious than reuse)	4
59	tech.	knowledge unstructured	4
15	DSE, tech.	dispersed teams have their own sharing environments	4
39	social	fear of knowledge parasites ("free-ride")	4
4	process-rel.	lack of knowledge distillation (experience engineering)	3
32	org., social	mistakes are tabooed and not recorded	3
33	org., social	lack of social network	3

58	tech.	system too slow	3
25	process-rel., org.	communication flows are restricted into certain directions (top-down)	3
7	DSE, org.	system is external and used as control mechanism	2
12	DSE, social	poor verbal / written communication skills	2
14	DSE, social	shared language not mastered well enough	2
16	DSE, tech., org.	no access to system	2
23	process-rel.	experiences are ambiguous: solutions can work in a situation and not work in another	2
28	org.	competitive internal work environment	2
29	org.	business units too big and knowledge sharing unmanageable	2
34	org., tech.	strong hierarchy, position-based status, centralized	2
35	org., tech.	not enough training and user support	2
46	social	differences of education levels	2
47	social	knowledge is too tacit to be captured and categorized (too time consuming to verbalize knowledge)	2
48	social	knowledge sharing is "bragging"	2
55	tech.	KMS too technology based	2
57	tech.	instability, system not reliable, not mature	2
62	tech.	low bandwidth to KMS	2
8	DSE, org.	Fear / uncertainty about access to knowledge base after project ends	1
10	DSE, social	do not trust others' knowledge	1
11	DSE, social	do not know where and how to submit knowledge	1
13	DSE, social	different national culture and ethnic background	1
19	process-rel.	content quality is low	1
27	org.	physical work environment restrict effective sharing practices	1
31	org.	no proper rollout strategy, lack of pilot phase	1
38	social	resistance to be known as expert	1
44	social	difference in experience	1
45	social	age differences	1

49	social, tech.	system is too complex	1
60	tech.	not searchable	1
61	tech.	no context in knowledge / documents	1

B. Heavy-weight Method Instantiations

This appendix presents short descriptions about methods that were classified as heavy-weight. The descriptions only name the tasks of the methods. The estimated durations are presented in Appendix C and the cognitive load in Appendix D. Besides a description, this appendix includes the sources and rationale why these methods are classified as heavy-weight, as well as literature for further details on each method.

B.1. Post-Mortem Experience Workshop

B.1.1. Project History Day

Project History Day was conducted by Apple and consists of 5 steps:

1. Distribute an anonymous project survey to project participants.

2. Collect objective information to determine objectives for the post-mortem workshop. Experience engineers or managers (I assume) collect project data, which should later provide information about subsequent scheduling efforts and help focus discussion during the workshop.

3. Conduct a debriefing meeting. All team members can give direct feedback about the project.

4. Participate in the Project History Day. The meeting itself should contain a limited number of people. During the meeting, root causes are created.

5. Publishing the results. The leadership team summarizes the workshop results in a report as an "Open Letter to Project Teams".

Claimed as heavy-weight by: Myllyaho et al. [185] call the method a large post-mortem analysis method, which strongly indicates that the method is heavy-weight. Dingsøyr et al. [87] mentions the method as one requiring much effort. The authors themselves mention that the method is "mentally exhausting" [62].

Further refer to: [62]

B.1.2. Learning Histories

The method consists of 6 steps:

1. Planning: Voluntary "champions" are assembled to discuss about "noticeable results" of the project.

2. Reflective interviews: The team of learning historians conducts interviews to gather points of view from a wide range of project participants. The interviewees should not analyze but just describe what happened.

3. Distillation: A small group of internal staff members create a report from the gathered data. The report has a special form. It should have a theme that captures the significance of an event. The themes should have compelling titles.

4. Writing: The team and the historians write a report together. This report should have a special two-column form and tell narratives from interviews on the one side and their essence on the other.

5. Validation: The organization has to validate the document. In addition, workshops are conducted consisting of the interviewees.

6. Dissemination: Experience are disseminated through workshops.

Claimed as heavy-weight by: Schindler and Eppler name the method a "heavy-weight concept" because of much required effort [210].
Further refer to: [210, 204]

B.1.3. Post-Project Review

The Post-Project Review (PPR) is a meeting, where the whole team as well as the project manager or a moderator take part. There are slightly different ways to set up and conduct a PPR. An overview is given by Koners and Goffin [155]. The objective of a PPR is to learn from mistakes and successes. Most often a PPR is conducted after a product has been introduced to market. The discussion method ranges from unstructured to a structured discussion following a guideline.

Claimed as heavy-weight by: Von Zedtwitz [262] reports on problems of PPRs. His research indicates that the reason for not conducting PPR often was lack of time "to complete a formal lessons-learned process". This indicates that the method requires a lot of effort. Harrison mentions the outcome of a PPR as "expensive wisdom" [116], which further indicates that this method takes long and requires much time.

Further refer to: [116, 53, 262, 155]

B.2. Experience Authoring

B.2.1. Postmortem Report

The main task of this method is for teams to write a document of 10 to 100 pages, which takes 3-6 months. They contain descriptions of what worked well or not so well in the project and what should be improved in the next project. Functional managers prepare a draft that is edited by the teams. Besides writing, the teams conduct a post-mortem meeting after each milestone to make midcourse corrections.

Claimed as heavy-weight by: Dingsøyr et al. [87] mention the method as one requiring much personal effort.

Further refer to: [71, 70, 69]

B.2.2. Write experience in Wiki

The experience bearer is provided a plain Wiki without any plugins, except a simple WYSIWYG editor.

According to own experience in a distributed collaborative project [179], installing a Wiki is a popular solution for a simple and "light-weight" EM system. Including this method should make apparent how light-weight or not so light-weight this method is.

Claimed as heavy-weight by: There is not direct source that claims this method as heavy-weight. However, according to many sources, Wikis alone as tools to contribute experience are often not accepted [252, 164, 80, 43, 128, 179, 113, 57, 202].

Further refer to: [161, 34].

B.2.3. Answer each question directly (without Answer Garden)

This method describes a setting where an expert is frequently asked for answer without Answer Garden. An expert would have to answer each question personally.

Claimed as heavy-weight by: There is no direct source that claims this method as heavy-weight. This method should make apparent how much effort is reduced by a concept like Answer Garden.

Further refer to: See description of Answer Garden [10, 12, 11].

B.2.4. Quality Patterns

Quality Patterns is a structure to package experiences to make them reusable. An experience should contain a classification, abstract, problem, solution, context, example, explanation, link to related experience and some administrative information. The explanation should contain a GQM [244] model.

Claimed as heavy-weight by: Schneider claims that this approach is heavy-weight [212].

Further refer to: [130]

C. Measurement of the Time to Share

This appendix presents details on the measurement of the time to share an experience used to create the portfolio (Figure 5.5). Figure C.1 presents the distribution of the the time to share an experience for light-weight and heavy-weight experience collection methods that were analyzed in this thesis. Table C.1 displays the times of the examined light-weight and Table C.2 of heavy-weight collection methods. Section C.1 explains the rationale behind the values.

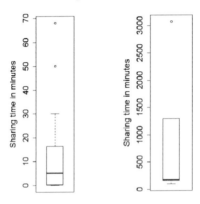

Figure C.1.: Distribution of the time to share for light-weight (left) and heavy-weight (right) methods. The time to share of *Postmortem Review* (HA3) is an extreme outlier and is not displayed in this plot.

Table C.1.: Time to share one experience by one experience bearer during a 2 years project.

Category	Method M	Steps and $t_{i,M}$	f	e	p	t_M
by-product	CRMT	(i) open tool and connect : 30sec.	4	40	5	15sec
	FOCUS	(i) press "instrument" button: 2sec.; (ii) start recording: 2sec.; (iii) start monitoring: 2sec.; (iv) save: 2sec.	2	20	1	0.8sec
data mining	e^3	no action	nr	nr	nr	0sec
	Autom. Rationale Extr.	no action	nr	nr	nr	0sec
	Delta Analysis	(i) participate a followup semi-struct. interview: 30min.	5	25	1	6min
interview	Reflective Guides	(i) answer one question in questionnaire: 10min	5	5	1	(10min)
		(ii) workshop duration: 1.5h	5	500	5	10min + 4.5min = 14.5min
exp. workshop	eWorkshop	(i) (i) read pre-meeting info sheet: 3min, (ii) memorize key points: 7min	1	0	1	(10min)
		(ii) participate meeting: 2h	1	533	15	10min + 3.38min = 13.38min
	LIDs	workshop duration: 1.5h	8	944	5	3.8min
	LPMR	workshop duration: 4h	1	24	5	50min
	Answer Garden	(i) open tool: 6sec.; (ii) answer a question, i.e. write a medium text of half a page: 20min.	176	176	1	20min
	Dandelion	(i) click on "reply message": 2sec; (ii) write short experience: 5min; (iii) send email: 2sec	150	150	1	5min
exp. authoring	Observation Sheet	(i) write short experience: 5min	150	150	1	5min
	Mail2Wiki A (contribute part of email with plugin)	(i) select part of email (text): 2sec; (ii) select one of the recommended pages or a previously saved as favorite page: 20sec; (iii) drag into selected page: 2sec.; (iv) edit email content (couple of sentences): 5min; (v) save: 2sec	150	150	1	5.26min
	Mail2Wiki B (contribute batch of emails without plugin)	(i) select email in email client: 2sec; (ii) drag into plugin panel at the side of email client: 2sec; (iii) write name of newly created page: 20sec.	150	150	1	24sec
	Mail2Wiki C (contribute batch of emails with plugin)	(i) select email as attachment: 5sec; (ii) type specific email address with tag: 5sec; (iii) send email:2sec	150	150	1	12sec
	Wiquila	(i) open browser: 6sec; (ii) open Wiki URL: 5sec; (iii) type short experience: 5min; (iv) submit: 2sec	150	150	1	5min

nr = not relevant

Table C.2.: Time to share one experience by one experience bearer during a 2 years project for heavy weight collection methods.

Category	Method M	Steps and $t_{i,M}$	f	e	p	t_M
exp. workshop	Learning Histories	(i) planning (conversation with "champions" about "notice-able results" [204]): 1.5h	1	0	3	(4.5h)
		(ii) reflective interview: 1h	1	20	1	(3min + 4.5h = 273min)
		(iii) write learning histories: 2800min	1	70	5	2800 + 273 = 3073min
	Post Project Review	(i) preparation for meeting: 30min; (ii) review meeting: 3h	1	11	5	96min
	Project History Day	(i) fill out electronic project survey: 1h		32	1	(1.9min)
		(ii) debriefing meeting: 1.5h	12	300	150	(9h)
		(iii) preparation (review private stores of status reports, emails, meeting minutes,...): 3h; (iv) conduct workshop: 6h	1	20	8	3.6h + 1.9min + 9h = 21.6h
	Post Project Review	(i) preparation for meeting: 30min; (ii) review meeting: 3h	1	11	5	96min
	Write experience in Wiki	(i) open browser: 6sec; (ii) open Wiki page: 2sec; (iii) login: 5sec.; (iv) write experience: 7min	155	155	1	7min
	Answer each question directly (without *Answer Garden*)	see *Answer Garden*	1552	176	1	176.36min
exp. authoring	Quality Patterns	(i) write one sheet [130]: 45min; (ii) think about restructuring experience: 15min; (iii) create GQM tree: 1h; (iv) write data and interpretation as explanation: 45min	150	150	1	135min
	Postmortem Report (Microsoft)	(i) write post-mortem document: 2.25 months	1	100	450	10.13 months
		(ii) hold post-mortem meeting: 1.5h	5	0	1	1.5h + 10.13months ≈ 10.13 months

nr = not relevant

Time to share in minutes (logarithmic)

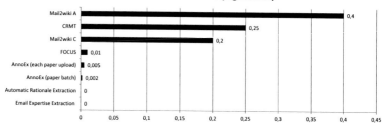

Figure C.2.: Time to share of one experience (only light-weight).

Time to share in minutes (logarithmic)

Figure C.3.: Time to share of one experience.

C.1. Rationale: Measurement of the Time to Share

The following listing presents the rationale behind the time to share estimations for each method.

CRMT

$t_{i,M}$: No information in literature [215] about duration and specific steps. Assume that a connection with remote participants is needed. Value is an educated guess.

f : Value is an assumption. Paper states that risk assessment must happen often.

e : Value is an assumption based on own experience. Assume that a prototype presentation will not yield too much rationale, which can be 10 experiences during a session. Considering $f = 4$, the sum of gained experiences is $10 \cdot 4 = 40$.

p : Literature did not provide any information on this. I Assume the default team size of 5.

FOCUS

$t_{i,M}$: Steps are named in [215] but not step time. Step durations are values from Table 5.3. Assume that no tool opening needed. Being an Eclipse plugin, Eclipse expected to be open for presentation.

f : No information in the paper. Assume two prototype meetings during a project.

e : Value is an assumption based on own experience. Assume that a prototype presentation will not yield too much rationale, which can be 10 experiences in a session. Considering $f = 2$, the sum of gained experiences is $10 \cdot 2 = 20$.

p : Assume that one developer presents his prototype including rationale.

Email Expertise Extraction (e^3)

$t_{i,M}$: Nothing to do for experience bearers, as "topics are generated through unsupervised clustering of message content" [51].

As experience bearers spend no effort, the rest of the values is not relevant.

Autom. Rationale Extraction

$t_{i,M}$: Rationale is extracted without any additional work by experience bearers.

As experience bearers spend no effort, the rest of the values is not relevant.

Delta Analysis

$t_{i,M}$: Delta identification is done by experience engineers. Following focused interviews are named but no duration given. Assume default duration of a semi-structured interview according to Table 5.3.

f : No information about frequency. Assume once in each development phase (requirements, design, implementation, testing, integration).

e : No information given. Assume not much (about 5): though many diffs may be found, most will not lead to interesting experiences. During the whole project, the number of collected experiences would be $5 \cdot 5 = 25$, considering $f = 5$.

p : Assume that interviews are mostly conducted with one person.

Reflective Guides

$t_{i,M}$: The duration to fill out one question is not given in the literature. I assume 10min per question according to own experience (some questions must be sophisticated). For the workshop duration no information is provided, so assume the default duration from Table 5.3.

f : The literature states that reflective guides should be conducted in each major development phase [170]. I assume a standard Waterfall process with 5 phases (requirements engineering, design, coding, testing, integration).

e : Questionnaire: As a result of the questionnaire, I assume that each question contains one experience. Workshop: As a result of the following workshop, I assume the sum of each questionnaire. I assume that one questionnaire contains 20 questions. Assuming 5 people in a team (see below) and $f = 5$, it makes altogether $5 \cdot 5 \cdot 20 = 500$ experiences throughout the project. I also assume that each questionnaire contains different experiences.

p : No information provided. I assume 5 people in a team.

eWorkshop

$t_{i,M}$: The name of the "pre-meeting information *sheet*" indicates that it is not very long. As no information is given, I assume it is one page long. The duration is taken from Table 5.3. The meeting duration is given in the literature [32].

f : The literature gives no hint on how often eWorkshop should be held. The paper reports on 3 workshops but only because the discussed problem required 3 workshops. This number depends on the problem to discuss. Thus, I assume this experience workshop as a post-mortem workshop.

e : The number of elicited experiences is given in the literature [32]. Reading the pre-meeting information sheet does not result in collected experiences.

p : Each person has the effort to read the pre-meeting information sheet. The list of meeting participants is given [32].

LIDs

$t_{i,M}$: The value is based on repeated experience in student software projects (3 months), where LIDs are conducted post-mortem. They last 1.5h. As LIDs should be conducted after each major phase [212] i.e. every couple of months, the student LID and its duration can be considered quite realistic.

f : I assume that LIDs are conducted every three months.

e : I analyzed 34 student LIDs from the last two software projects in the years 2012/13 and 2013/14. The analysis revealed that a LID contains 118 experiences on average (see Figure C.4). As a heuristic for experience identification, I assumed that each paragraph

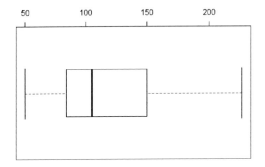

Figure C.4.: Distribution of the number of identified experiences in 34 LID documents.

(text separated by a line break) contains one statement and reflects one experience. I did not count mere naming of activities (often as titles in the chronological recollection, e.g. "1st customer meeting") as an experience. The titles represent experience context. Altogether, LIDs produce $8 \cdot 118 = 944$ experiences during a 2 years project, considering $f = 8$.

p : Assume one default team.

LPMR

$t_{i,M}$: The duration is named as half a (work) day [87], which is 4h.

f : This is a post-mortem workshop.

e : This is the average of 21 and 28, which are the found experience items in the study [87].

p : The study mentions one team of 5-10 people participating [87].

Project History Day

$t_{i,M}$: Time estimation for project survey provided in the literature [62]. Time for debriefing was not given. I assume the default value from Table 5.3. The preparation activity is named but the time is not given in the literature. I assume 3h, as it seems to be a fairly time consuming task to go through the whole store of status reports, emails and meeting minutes. The duration of the workshop itself is provided.

f : Survey: It is not clear in the literature [62], if the surveys are distributed once or more times. Thus I assume the minimum, i.e. once. Review: This is a post-mortem review. For preparation and the workshop, both activities are done once.

e : Questionnaire: According to Collier et al. [62], a questionnaire contains a dozen questions in each of 8 categories. This means approximately $12 \cdot 8 = 96$ questions altogether. I also assume that $1/3$ of the answers contain experiences. This results in $96/3 = 32$ experiences. Debriefing: Collier et al. state that "[a] project team of 150 memebers who hold a dozen debriefing meetings can produce hundreds of flip-chart pages of issues, comments, and solutions" [62]. Estimating "hundreds of slides" as about 300 slides and assuming that each slide holds one experience, I estimate 300 experiences altogether during the project. Workshop: In the case study [62] the participating team selected 20 root-cause relationships to discuss during the workshop. I assume that each root-cause is based on at least one experience.

p : Questionnaire: This is always filled out by one person each. Debriefing and workshop: The participant number is given in the literature. [62].

Postmortem Report (Microsoft)

$t_{i,M}$: I assume a report writing process, where a lead group writes a sketch, then team members comment on the sketch and the lead team edits the document. The literature mentions that "groups generally take three to six months to put a post-mortem document together" [71]. There is no information if the team works full time on the document or not. Usually, software engineers work on multiple projects. I assume that they work $1/2$ of the time on the document. Further I assume the average of 3 to 6 months, which is 4.5 months. Working half time means that the team spends $4.5/2 = 2,25$ months of actual work on writing the document. To estimate the time in minutes, I assume that a work month has 21 days (These were the average work days in Lower Saxony in 2013 [1]) and one day contains 8 work hours. This results in $2.25 \cdot 21 \cdot 8 \cdot 60 = 22680$min. For the post-mortem meeting

---

[1]http://www.schnelle-online.info/Arbeitstage/Anzahl-Arbeitstage-2013.html, accessed May 11 2014

no timeframe is named. Therefore, I assume a default duration from Table 5.3.

f : The report is written only once. The post-mortem meeting should be conducted after each milestone [71, p. 332]. I assume a Waterfall development process with 5 milestones.

e : According to Cusumano and Selby a post-mortem report has 10-100 pages [71]. I assume that each page has one recommendation. The post-mortem meetings have the goal to discuss the results of the last post-mortem [71]. As a simplified estimation, I assume that during such a meeting no new experiences are collected, but the old ones are engineered into optimized schedule or midcourse corrections.

p : Cusumano [69] names project size of a Microsoft project that was mentioned in relation to creating post mortem reports in [71]. Further I assume that the whole team responsible in creating the report. For the meeting, the effort concerns only one person, since no new experiences are shared.

Learning Histories

$t_{i,M}$: Planning: No time given in literature. I assume the default value from Table 5.3. Interview: No time is given either. However, according to description of the interview process [204], participants should tell what happened, tell a story. This is similar to how *LIDs* are conducted. There students tell what happened in chronological order. This takes approximately 30 to 40min according to own experience in with student LIDs. Considering a 2-years project, this will probably take longer, as more happened during the time. However, according to [204], interviewers ("learning historians") present an outline of "noticeable results" to the interviewees as anchor points to talk about. This should keep the interview time at bay. Writing: The writing duration is not given, only the fact that the whole team writes the report [204]. According to Schindler and Eppler [210] a report can be between 20 and 100 pages long. Assuming the average of 70 pages, a default writing duration would be $70 \cdot 40\text{min} = 2800\text{min}$.

f : Planning: no information given. I assume that an identification is done only once. Interview: no information given. I assume that each person is interviewed only once. Writing: This is done only once.

e : Planning: According to Roth [204] this meeting identifies results, *not* experiences. The paper does not state that experiences are identified. Interview: no information or indicator given. I assume that the outcome is similar to LIDs but a bit less, since the participants have less time. A LIDs session produces 118 experiences from 5 experience bearers. Taking a slightly smaller number, an approximation would be $100/5 = 20$ experiences for one person. Writing: A report can be between 20 and 100 pages long. I assume the average of 70 pages and assume that one page contains one experience. This is roughly compatible with the example excerpt by Roth and Kleiner [204].

p : Planning: no information or hint given. I assume that managers or project leaders take part in this planning conversation, as they are usually in charge to oversee and report project results. I assume a low participant number. Interview: Interview is assumed to be conducted with one interviewee. Writing: Besides the knowledge that the team writes the report, no information on the team size is given. I assume a default team size of 5 people.

Post Project Review (PPR)

$t_{i,M}$: Preparation: The case study does not present a duration of the preparation. Mostly, the team prepares positive and negative points to discuss. I assume a preparation time of 30min. Review: Koners and Goffin [155] present an overview of several PPR variants in different companies. The duration ranges from 3h, full day (8h), 2h, 1h, 2h. I take the average of these values: $16h/5 = 3.2h \approx 3h$.

f : This is a post-mortem workshop.

e : The only indicator to the experience number is the metaphor count in the case study [155]: "Overall, 55 metaphors and stories were identified across the 15 hours of PPRs observed: on average, one metaphor every 20 minutes." This results in $15h/3h = 5$ PPR meetings leading to $55/5 = 11$ metaphors in one meeting on average. I assume that one metaphor describes one experience.

p : According to Koners and Goffin [155] one team takes part.

Answer Garden

$t_{i,M}$: I assume that the expert already knows the answer and only consider writing time. The time to search for answer is considered a constant, as the expert would have to research anyway with or without Answer Garden. The reason for assuming that an expert would write only a medium text is own experience with FAQ systems, which are usually between couple of sentences and half a page.

f : An evaluation of Answer Garden (AG) [11] reports that each expert had to answer 2 questions per week. In Germany, people have 88 work weeks in 2 years (52 weeks $-$ 2 weeks holidays (11 days) $-$ 6 weeks vacation = 44 weeks). Extrapolating, this means $44 \cdot 2 \cdot 2 = 176$ answers.

e : I assume that each answer is one experience.

p : Each expert writes an answer personally.

Dandelion

$t_{i,M}$: I assume the email-version of Dandelion and that the default email client is already open. I further assume a situation, where the experience bearer is responding to email inviting him to be co-author. I do not consider the effort of the coordinator. The latter can be considered an experience engineer. Though in the evaluation [58] people created "short" documents of 1000 words, these documents were no experiences. I assume a use case for Dandelion where experience bearers would write short, ah hoc experiences like observation sheets [231].

f : See Observation Sheet.

e : See Observation Sheet.

p : The author is a single person.

Observation Sheet

$t_{i,M}$: Writing a short experience on a sheet of paper takes about 5min according to Table 5.3.

f : The experience bearer writes only one experience on one sheet.

e : The experience number is an extrapolation based on the results of a 5 weeks global development project [230]. The project resulted in 44 observation sheets [23] authored by 5 participants. On average, one person contributed $44/5 = 8.8$ experiences during 5 weeks. Two years consist of 88 weeks. Extrapolating, and assuming a constant and continuous writing amount throughout the project duration, one person would share $88/5 \cdot 8.8 = 149.6$ experiences.

p : The author is a single person.

Mail2Wiki A (contribute part of an email with the plugin)

$t_{i,M}$: The steps are described by Convertino [65]. The durations are taken from Table 5.3. Here, I do not count the time to decide which text to select. This is equal to the time to decide to write down an experience, which is not calculated either. The confidence is low because I do not know the estimated time for the decision which page to choose. This is a guess, which highly depends on the number of recommendations. I assume though, there should be no more than 2-3 categories, thus scanning them and deciding should not take long.

f : See Observation Sheet.

e : See Observation Sheet.

p : The author is a single person.

Mail2Wiki B (contribute one or more emails without the plugin)

$t_{i,M}$: I assume Mail2Wiki as a tool for writing observation sheets [231], i.e. writing short experiences. The steps are named in the literature [65]. I assume a worst case where the user does not choose a recommended page to drag emails into. In this case, a new Wiki page is created and he has to spend time naming the page. The values are taken from Table 5.3.

f : See Observation Sheet.

e : See Observation Sheet.

p : The author is a single person.

Mail2Wiki C (contribute a batch of emails with the plugin)

$t_{i,M}$: The steps are described by Convertino [65]. The durations are taken from Table 5.3.

f : See Observation Sheet.

e : See Observation Sheet.

p : The author is a single person.

Wiquila

$t_{i,M}$: I assume Wiquila as a tool for writing observation sheets [231], i.e. writing short experiences.

f : See Observation Sheet.

e : See Observation Sheet.

p : The author is a single person.

Write experience in Wiki

$t_{i,M}$: The duration is based on the values in Table 5.3. I assume that the Wiki URL is bookmarked and can be opened with one click. I assume that there is no ready form to enter the experience. The experience bearer has to create a new Wiki page. The writing time of the experience is assumed a bit higher than with e.g. *Observation Sheet* (5min), since writing in a Wiki even with a simple WYSIWYG[2] editor (like the one in the current Wikipedia MediaWiki 1.22) requires more interaction (e.g. for adding links or pictures) and recall of the Wiki syntax. The latter is often needed despite the WYSIWYG editor for more sophisticated text or to create semantic relations (e.g. assigning categories). I assume

---

[2]What You See Is What You Get

184

that for a short experience, the author would need about 7min on average to write with a MediaWiki.

f : See Observation Sheet.

e : See Observation Sheet.

p : The author is a single person.

Answer each question directly (without Answer Garden)

$t_{i,M}$: See Answer Garden.

f : Without the Answer Garden question delegation system, an expert must answer each question personally. According to Ackerman [11] Answer Garden was visited 194 times during 3 months. Assuming a constant usage, frequency can be extrapolated to $194 \cdot 8 = 1552$ times.

e : See Answer Garden

p : See Answer Garden.

Quality Patterns

$t_{i,M}$: I assume that the experience bearer himself has to write the experience according to the given pyramide pattern. No time frames are given in the literature [130], thus I assume default values from Table 5.3 for writing times. For step (iv) I assume that the explanation would have approximately the length of one page, indicated by the lengthy example in the literature [130]. Besides writing, I assume that the experience bearer will have to think about how to adapt his experience to the given pattern. In comparison to e.g. Answer Garden, where I assume that the expert already knows what to answer and do not consider the time to think about the answer, here, the experience bearer already knows what to write, but has to adapt his writing style to fit the structure.

f : Similar to Observation Sheet.

e : I assume that one sheet contains one experience.

p : The author is a single person.

D. Measurement of Cognitive Load

This appendix presents details on the measurement of cognitive load. The boxplots in Figure D.1 present the cognitive load measurement result of the analyzed experience collection methods, while Figure D.2 presents the measurement results in a descending order. Table D.1 displays the values for each cognitive load factor. Section D.1 presents rationale to explain the choice of values.

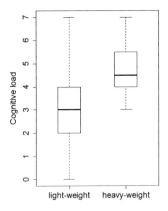

Figure D.1.: Cognitive load distribution.

D.1. Rationale: Measurement of Cognitive Load

FOCUS

Concepts or rules to recall : Nothing to recall. Experience bearer follows the buttons on the panel [215].

Mental demand of a task : No additional tasks to perform.

Integration into present tooling : FOCUS is integrated into Eclipse, a widespread IDE [103].

Table D.1.: Cognitive load of all examined collection methods. The gray shaded methods are classified as heavy-weight.

Category	Method M	Recall	Mental demand	Tool integr.	Process integr.	Ad hoc sharing	Cognitive load
by-product	FOCUS	0	0	0	0	1	1
	CRMT	0	1	1	0	1	3
	e^3	0	0	0	0	0	0
data mining	Autom. Rationale Extr.	0	0	0	0	0	0
	Delta Analysis	0	0	0	1	1	2
interview	Reflective Guides (questionnaire/workshop)	0/0	2/0	0/0	1/2	2/1	5 (/3)
	eWorkshop (pre-meeting info sheet/meeting)	0/0	2/1	0/0	2/2	2/1	6 (/4)
	LIDs	0	0	0	2	1	3
exp. workshop	LPMR	0	1	0	2	1	4
	Project History Day	0	2	0	2	2	6
	Learning Histories (conversation/interview/writing)	0/0/0	1/0/2	0/0/0	2/2/2	1/1/2	(4/3/) 6
	Post Project Review preparation/meeting	0/0	2/2	0/0	2/2	0/1	(4/) 5
	Answer Garden	0	1	0	1	2	4
	Dandelion	1	1	0	1	0	3
	Observation Sheet	0	1	0	1	0	2
	Mail2Wiki A (part of email with plugin)	0	1	0	1	0	2
	Mail2Wiki B (batch of emails without plugin)	1	0	0	1	2	4
exp. authoring	Mail2Wiki C (batch of emails with plugin)	0	0	0	1	2	3
	Wiquila	0	1	1	2	0	4
	Answer each question directly (without *Answer Garden*)	0	1	0	1	2	4
	Postmortem Report	0	2	2	0	2	6
	Quality Patterns	2	2	1	2	0	7
	Write experience in a Wiki (without plugins)	2	1	2	2	0	7

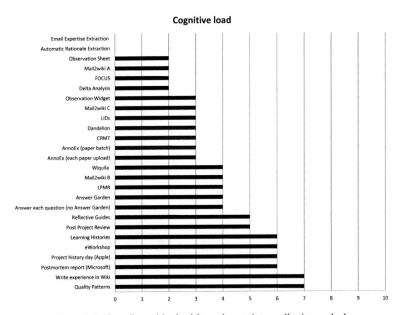

Figure D.2.: Overall cognitive load for each experience collection method.

Integration into working processes : Refer to principle 2 of the method category *experience as by-product* (Section 4.3.1).

Ability of ad hoc externalization : Experiences (rationale) can be shared ad hoc only during the prototype presentation.

CRMT

Concepts or rules to recall : Nothing to recall.

Mental demand of a task : Participants have to read others' messages in the chat. This communication channel (chat) has a higher latency and channel bandwidth than a face-to-face meeting [229, p. 111].

Integration into present tooling : According to Schneider [215] the given implementation is a standalone tool, which has a simple user interface. I assume that it behaves like a common chat tool. The risk portfolio view should be also well known to risk analysts.

Integration into working processes : Refer to principle 2 of the method category *experience as by-product* (Section 4.3.1.

Ability of ad hoc externalization : Experiences (rationale) can be shared ad hoc only during the risk analysis meeting.

Email Expertise Extraction (e^3)

There is nothing to do for the experience bearer.

Autom. Rationale Extraction

There is nothing to do for the experience bearer.

Delta Analysis

The task that can be cognitively demanding is the focused interview after "delta" identification.

Concepts or rules to recall : Nothing to recall.

Mental demand of a task : I assume that the interviewee does not read or write anything during the interview or does analytic thinking. He just answers questions.

Integration into present tooling : No new tool introduced.

Integration into working processes : The interview is a separate appointment and outside of the usual work process.

Ability of ad hoc externalization : Experiences can be shared ad hoc only during the interview.

Reflective Guides

Concepts or rules to recall : Questionnaire: Nothing to recall. Workshop: Literature [170] does not specifically mention any workshop method that would be required to recall.

Mental demand of a task : Questionnaire: According to an excerpt of questions [170], they may require a lot of (analytical) thinking. Workshop: Simple discussion on experiences collected in the questionnaires.

Integration into present tooling : Questionnaire: No special tool needed, just pen and paper. Workshop: no tool needed.

Integration into working processes : Questionnaire: Filling out the questionnaire is an extra task. Workshop: This is a separate appointment and outside of the usual work process.

Ability of ad hoc externalization : Experiences can be shared ad hoc only during the workshop.

eWorkshop

Concepts or rules to recall : Nothing to recall.

Mental demand of a task : Pre-meeting info sheet: reading the sheet. Assuming a long sheet (> 1 page) the mental demand is high. Workshop: this is a chat, see *CRMT*.

Integration into present tooling : No special tools needed. Literature emphasizes on "us[ing] simple collaboration tools, thus minimizing potential technical problems and decreasing the time it would take to learn the tools" [32].

Integration into working processes : All tasks and activities are separate and not part of a usual development process.

Ability of ad hoc externalization : Pre-meeting info sheet: no externalization, only internalization. Workshop: Experiences can be shared ad hoc only during the workshop.

LIDs

Concepts or rules to recall : Nothing to recall.

Mental demand of a task : No mental demand. The participants do not have to read or write anything, just tell their story.

Integration into present tooling : No new tooling needed. The meeting minutes are written with a usual word processing software.

Integration into working processes : This is a separate appointment and outside of the usual work process.

Ability of ad hoc externalization : Experiences can be shared ad hoc only during the workshop.

LPMR

Concepts or rules to recall : Nothing to recall.

Mental demand of a task : During the workshop, participants create a root cause diagram. This may require some analytic thinking.

Integration into present tooling : No special tools needed.

Integration into working processes : This is a separate appointment and outside of the usual work process.

Ability of ad hoc externalization : Experiences can be shared ad hoc only during the workshop.

Project history Day

Concepts or rules to recall : Nothing to recall.

Mental demand of a task : "Project History Days are psychologically and mentally exhausting" [62].

Integration into present tooling : No special tools needed.

Integration into working processes : This is a separate appointment and outside of the usual work process.

Ability of ad hoc externalization : Experiences can be shared ad hoc only during the workshop.

Postmortem Report

Concepts or rules to recall : Nothing to recall.

Mental demand of a task : The experience bearer is required to write a very long document.

Integration into present tooling : No special tools needed.

Integration into working processes : This is a separate appointment and outside of the usual work process.

Ability of ad hoc externalization : Experiences cannot be shared ad hoc because they are written down. There is no reflection-in-action.

Learning Histories

Concepts or rules to recall : Nothing to recall.

Mental demand of a task : Conversation: The participants have to decide what significant results are. Interview: No analytical thinking. The participant just recollect the events. Writing: According to Roth and Kleiner writing the report "places an enormous burden on the artistry of the editor" [204]. Besides, the document is very long.

Integration into present tooling : No special tools needed.

Integration into working processes : This is a separate appointment or process and outside of the usual work process.

Ability of ad hoc externalization : Workshops: Experiences can be shared ad hoc only during the interview. Writing: Experiences can only be written down at a fixed time.

Post Project Review

Concepts or rules to recall : No specific rules to recall.

Mental demand of a task : Preparation: Participants need to read many documents or write positive and negative aspects or reflect analytically [155]. Meeting: Analytical thinking required to create causes and consequences.

Integration into present tooling : No special tools needed.

Integration into working processes : This is a separate appointment and outside of the usual work process.

Ability of ad hoc externalization : Preparation: no ad hoc sharing, as no experience sharing happens. Meeting: Experiences can be shared ad hoc only during the meeting.

Answer Garden

Concepts or rules to recall : Nothing to recall.

Mental demand of a task : Write a medium text.

Integration into present tooling : The expert answers by email [10].

Integration into working processes : The expert has to write the answer (a medium sized text).

Ability of ad hoc externalization : Ad hoc sharing not possible as answers are only triggered by questions.

Dandelion

Concepts or rules to recall : The method requires filling out an email template containing special characters and sections. The author has to recall where to write which information.

Mental demand of a task : The author is assumed to write a short text.

Integration into present tooling : Dandelion is an email client plugin.

Integration into working processes : Filling out an email template is a slight change of the usual email writing process.

Ability of ad hoc externalization : Assuming that the experience bearer constantly works at his desk and PC, he can share his experiences anytime.

Observation Sheet

Concepts or rules to recall : Nothing to recall.

Mental demand of a task : The experience bearer has to write a short text.

Integration into present tooling : No tooling needed, only pen and paper.

Integration into working processes : The experience bearer still needs to disrupt a process to explicitly write down the experience. This is only a minor change in the usual workflow because it does not take long.

Ability of ad hoc externalization : The tool can be used anytime, assuming that a sheet of paper and pen are at hand.

Mail2Wiki A (contribute part of an email with the plugin)

Concepts or rules to recall : Nothing to recall.

Mental demand of a task : The experience bearer has to write a short text in the case where he writes free text experience.

Integration into present tooling : Mail2Wiki is a plugin of a well known email client (MS Outlook).

Integration into working processes : Copying and pasting a part of an email into the plugin is only a slight change of the usual email writing process [57].

Ability of ad hoc externalization : It is possible to write free text experiences instead of email anytime.

Mail2Wiki B (contribute one or more emails without the plugin)

Concepts or rules to recall : The experience bearer has to recall a special email address syntax.

Mental demand of a task : The experience bearer only forwards emails to a dedicated server.

Integration into present tooling : Mail2Wiki is a plugin of a well known email client (MS Outlook).

Integration into working processes : Forwarding emails is just a slight change of the usual email writing process of every day work. No adjustment needed.

Ability of ad hoc externalization : The experiences in the forwarded emails have already been externalized. No ad hoc sharing possible at the time of forwarding.

Mail2Wiki C (contribute a batch of emails with the plugin)

Concepts or rules to recall : Nothing to recall.

Mental demand of a task : Nothing to read or write. I assume that the emails to share are already chosen.

Integration into present tooling : Mail2Wiki is a plugin of a well known email client (MS Outlook).

Integration into working processes : Dragging and dropping a bunch of emails into the plugin is only a slight change of the usual email writing process [57].

Ability of ad hoc externalization : The experiences in the forwarded emails have already been externalized. No ad hoc sharing possible at the time of forwarding.

Wiquila

Concepts or rules to recall : Nothing to recall.

Mental demand of a task : The experience bearer has to write a short text.

Integration into present tooling : Though being a standalone tool, the WYSIWYG editor facilitates interaction with and authoring in a Wiki. Thus, the plugin enables writing in a Wiki like in a usual work processing software.

Integration into working processes : Writing the experience is an extra task.

Ability of ad hoc externalization : The tool can be used anytime assuming that the author is at the workplace.

Write experience into a Wiki (without plugins)

Concepts or rules to recall : The experience bearer has to recall the Wiki syntax.

Mental demand of a task : The experience bearer has to write a short text.

Integration into present tooling : I assume that the Wiki is newly introduced and has not been used in the company. There is much evidence that Wikis still suffer adoption problems in enterprises (e.g. [252, 164, 80, 43, 128]), also confirmed by own experience in a joint venture [179].

Integration into working processes : Writing the experience is an extra task.

Ability of ad hoc externalization : The tool can be used anytime assuming that the author is at the workplace.

Quality Patterns

Concepts or rules to recall : The experience bearer has to recall how to compose a GQM tree.

Mental demand of a task : The experience bearer has to write a long (about 1 page according to the example in the literature [130]) text.

Integration into present tooling : I assume that writing the experience would require some form. According to Houdek and Kempter [130] collected experiences are disseminated over the Web. They do not mention which medium is used to collect the experiences (paper or web browser). I assume a form in the web browser. As the browser can be considered "intuitive" and "part of [the] daily work" [130], this would mean only a slight change and adaptation to the new tooling.

Integration into working processes : Writing the experience is an extra task.

Ability of ad hoc externalization : The tool can be used anytime assuming that the author is at the workplace.

E. Rationale: Measurement of Organizational Effort

This appendix presents the rationale on the effort measure of the organizational effort for all examined experience collection methods.

E.0.1. FOCUS

Min. EE effort: The record already produces a benefit. No anonymization and filtering needed, since I assume it is an official presentation, where only non-classified information is shared with partners.

Admin. effort: Installation of the plugin and uploading the record is necessary.

E.0.2. CRMT

Min. EE effort: The record already produces a benefit. No anonymization and filtering needed, since I assume it is an official meeting, where only non-classified information is shared with partners.

Admin. effort: Installation of the tool and uploading of the chat records are needed.

E.0.3. Email Expertise Extraction (e^3)

Min. EE effort: I do not consider information on expertise as threat to any possible NDA policies. On the contrary, the goal of e^3 is to support lack of awareness about experts in a distributed project.

Admin. effort: The analysis tool must be installed.

E.0.4. Autom. Rationale Extraction

Min. EE effort: The experience engineer (EE) has to manually sort out "noise" from the extracted rationale, since the algorithm does not have a good precision (60.5%) and recall (24%) [199].

Admin. effort: Only installation of the extraction tool is needed.

E.0.5. Delta Analysis

Min. EE effort: The EE needs to go through the diffs (task 1) of documents and identify interesting "deltas" (task 2).

Admin. effort: I assume that presenting the deltas to the experience bearers (render them the immediate benefit) happens during the focused interviews, the EE must prepare, conduct it and upload the results.

E.0.6. AnnoEx

Min. EE effort: The uploaded annotations already render a benefit. The annotations and uploaded documents must be checked if they contain sensible information.

Admin. effort: The tool and Wiki must be installed and the Wiki configured.

E.0.7. Reflective Guides

Min. EE effort: EE has to create a summary document from the questionnaire results but does neither have to derive recommendations (task 7), nor identify experiences (task 2). I assume that the categories (task 5) are provided by the questionnaire categories.

Admin. effort: The EE must prepare and conduct the workshop.

E.0.8. eWorkshop

Min. EE effort: EE writes a summary of the discussion (chat).

Admin. effort: The EE must prepare and conduct the workshop.

E.0.9. LIDs

Min. EE effort: The EE must only filter and anonymize the LID.

Admin. effort: The EE must prepare and conduct the workshop. Afterwards he must properly upload the documents and create links to other documents.

E.0.10. LPMR

Min. EE effort: The EE has to write a report. The report is immediately beneficial because it is created the day after the meeting. The report is the only valuable asset that is accessible for other team members.

Admin. effort: The EE must prepare and conduct the workshop. Afterwards he must properly upload the documents and make them accessible to other teams.

E.0.11. Project History Day

Min. EE effort: The EE has to write a report.

Admin. effort: The EE must prepare and conduct the workshop. Afterwards he must properly upload the documents and make them accessible to other teams.

E.0.12. Learning Histories

Min. EE effort: The EE has to write a report.

Admin. effort: The EE must prepare and conduct the workshop. Afterwards he must properly upload the documents and make them accessible to other teams.

E.0.13. Post Project Review

Min. EE effort: The EE has to write a report.

Admin. effort: The EE must prepare and conduct the workshop. Afterwards he must properly upload the documents and make them accessible to other teams.

E.0.14. Answer Garden

Min. EE effort: I assume that the answers are already formulated in a desirable form. They only have to be filtered (they are already anonymized).

Admin. effort: The system has to be installed and the categories defined.

E.0.15. Dandelion

Min. EE effort: The written experiences must be checked if they contain sensible information.

Admin. effort: The tool has to be installed and the environment configured.

E.0.16. Observation Sheet

Min. EE effort: The written experiences must be checked if they contain sensible information.

Admin. effort: The paper sheets have to be at least scanned. Therefore an experience base (e.g. a Wiki) has to be installed and configured.

E.0.17. Mail2Wiki A (contribute part of an email with the plugin)

Min. EE effort: The written experiences must be checked if they contain sensible information.

Admin. effort: The tool and a Wiki have to be installed and the Wiki configured.

E.0.18. Mail2Wiki B (contribute one or more emails without the plugin)

Min. EE effort: The written experiences must be checked if they contain sensible information.
Admin. effort: The tool and a Wiki have to be installed and the Wiki configured.

E.0.19. Mail2Wiki C (contribute a batch of emails with the plugin)

Min. EE effort: The written experiences must be checked if they contain sensible information.
Admin. effort: The tool and a Wiki have to be installed and the Wiki configured.

E.0.20. Wiquila

Min. EE effort: The written experiences must be checked if they contain sensible information.
Admin. effort: The tool and a Wiki have to be installed and the Wiki configured.

E.0.21. Write experience into a Wiki (without plugins)

Min. EE effort: The written experiences must be checked if they contain sensible information.
Admin. effort: The Wiki has to be installed and the Wiki configured.

E.0.22. Answer each question directly (without *Answer Garden*)

Min. EE effort: The written experiences must be checked if they contain sensible information.
Admin. effort: The system has to be installed and the categories defined.

E.0.23. Postmortem Report

Min. EE effort: No EE tasks needed, as the report is created by experience bearers.
Admin. effort: No administrative tasks are known.

E.0.24. Quality Patterns

Min. EE effort: The written experiences must be checked if they contain sensible information.
Admin. effort: The web application has to be installed and configured (create the form).

F. Measurement of Experience Maturity and Rawness

Table F.1.: Maturity and rawness of all examined experience collection methods. The gray shaded methods are classified as heavy-weight.

Category	Code	Method	Maturity	Rawness
by-product	B1	CRMT	3	6
	B2	FOCUS	3	5
data mining	D1	e^3	2	1
	D2	Autom. Rationale Extr.	3	5
	D3	Delta Analysis	2	6
interview	I1	Reflective Guides	4	3
exp. workshop	W1	eWorkshop	4	6
	W2	LIDs	4	4
	W3	LPMR	4	3
	HW1	Project History Day	5	3
	HW2	Learning Histories	5	3
	HW3	Post Project Review	5	3
exp. authoring	A1	Observation Sheet	2	5
	A2	Answer Garden	3	3
	A3	Dandelion	2	5
	A4.1	Mail2Wiki A (part of email with plugin)	2	5
	A4.2	Mail2Wiki B (batch of emails without plugin)	2	6
	A4.3	Mail2Wiki C (batch of emails with plugin)	2	5
	A5	Wiquila	2	5
	HA1	Write experience into a Wiki	2	5
	HA2	Answer each question directly (without *Answer Garden*)	3	3
	HA3	Postmortem Report	5	3
	HA4	Quality Patterns	2	2

The boxplots in Figure F.1 represent the experience maturity and rawness distribution of the examined light-weight and heavy-weight experience collection methods (Table F.1). They show that light-weight collection methods have a lower average maturity and a higher rawness.

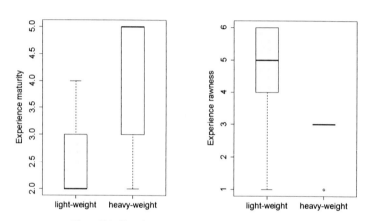

Figure F.1.: Experience maturity and rawness distribution.

G. Questionnaire for the Evaluation of OWHS

Code: _____

Fragebogen: Einfluss von Kritiken beim Schreiben von Erfahrungen

1. Bitte hilf uns zu verstehen, unter welchen Umständen du normalerweise Erfahrungen teilen würdest oder nicht.

	Stimme voll und ganz zu	Stimme eher zu	Stimme eher nicht zu	Stimme nicht zu	Keine Meinung
Zögerst du oft eine Erfahrung zu teilen, weil du nicht weißt, ob sie nicht von anderen als trivial empfunden wird?	○	○	○	○	○
Zögerst du oft eine Erfahrung zu teilen, weil du nicht weißt, wie du sie am besten formulieren sollst?	○	○	○	○	○
Wenn du etwas Relevantes gelernt hast, teilst du es gerne?	○	○	○	○	○
Wenn du etwas neu Gelerntes teilen möchtest, weißt du dann immer wo und mit wem?	○	○	○	○	○

2. Bitte hilf uns den Effekt von Kritiken bei der Dokumentation von Erfahrungen besser zu verstehen, indem du deine Übereinstimmung mit den folgenden Aussagen bewertest.

	Stimme voll und ganz zu	Stimme eher zu	Stimme eher nicht zu	Stimme nicht zu	Keine Meinung
Empfandst du die Kritiken als hilfreich beim Schreiben der Erfahrungen?	○	○	○	○	○
Empfandst du die Kritiken als berechtigt?	○	○	○	○	○
Fühlst du dich sicherer, wenn du Feedback durch die Kritiken erhältst?	○	○	○	○	○

	Stimme voll und ganz zu	Stimme eher zu	Stimme eher nicht zu	Stimme nicht zu	Keine Meinung
Findest du, dass die Kritiken die Hürde, eine neue Beobachtung einzugeben, senken?	○	○	○	○	○
Würden dir Erfahrungen, die mit Unterstützung von Kritiken erstellt wurden, mehr helfen als solche ohne Kritiken?	○	○	○	○	○
Würdest du mit einem Tool, das dich mit Kritiken unterstützt, mehr Erfahrungen schreiben?	○	○	○	○	○
Empfandst du die Kritiken als überwiegend störend in deinem Gedankengang?	○	○	○	○	○
Haben dir die Kritiken die Eingabe einer Erfahrung erschwert?	○	○	○	○	○
Empfandst du die Eingabe wegen der Kritiken insgesamt verlangsamt?	○	○	○	○	○

3. Hast du schon mal deine Erfahrungen in z.B. Online-Foren oder sozialen Netzwerken mitgeteilt?

Ja   Nein

○   ○

*Bei Antwort **Nein** weiter bei Frage 5.*

4. Wie oft teilst du deine Erfahrungen in z.B. Online-Foren oder sozialen Netzwerken mit?

Täglich	Wöchentlich	Monatlich	Seltener als Monatlich
○	○	○	○

5. Bitte hilf uns die Kritiken zu verbessern, indem du uns mitteilst, welche Kritiken du als hilfreich oder nicht hilfreich empfunden hast.

	Sehr hilfreich	Eher hilfreich	Weniger hilfreich	Gar nicht hilfreich	Keine Meinung/nicht aufgetaucht
✪ Die Beobachtung / Emotion / Schlussfolgerung enthält keinen Text.	○	○	○	○	○
⚠ Ihre Beobachtung / Schlussfolgerung ist sehr knapp.	○	○	○	○	○
⚠ Ihre Beobachtung / Schlussfolgerung könnte schwierig zu lesen sein.	○	○	○	○	○
⚠ Ihre Emotion wird möglicherweise nicht deutlich.	○	○	○	○	○
⚠ Ihre Schlussfolgerung enthält möglicherweise keine Empfehlung.	○	○	○	○	○
⚠ In Ihrer Schlussfolgerung fehlen möglicherweise Bezugspersonen oder Verantwortliche.	○	○	○	○	○
⚠ Ihre Beobachtung / Emotion / Schlussfolgerung enthält Akronyme.	○	○	○	○	○
› Der Autor ist nicht angegeben.	○	○	○	○	○
› Beobachtung ist...	○	○	○	○	○
› Emotion ist...	○	○	○	○	○
ⓘ Erfahrung besteht aus einer Beobachtung, Emotion und Schlussfolgerung.	○	○	○	○	○

6. Welche Kritiken fandst du besonders hilfreich? Bitte begründe kurz.

7. Welche Kritiken haben dich besonders gestört bzw. waren überhaupt nicht sinnvoll? Bitte begründe kurz.

8. Bitte sag uns, welche der 4 Erfahrungen für dich von guter und welche von schlechter Qualität sind, indem du den Erfahrungen nach ihrer Qualität (verständlich, hilfreich, informativ) einem Rang von 1-4 zuweist (beste mit 1, jede Zahl einmal).

Rang	Erfahrung
	Beobachtung: E-Meeting mit LifeSize ist problematisch. Nach 1h ist immer noch unklar, wie man das Tool bedienen soll, wie man Video bekommt und wie man mit Finnland und Russland verbinden soll. Die Finnen und Russen machen auch nichts auf ihrer Seite. **Emotion:** Das macht mich wütend! Warum müssen wir immer die Arbeit machen?! **Schlussfolgerung:** Alle Teams müssen sich besser vorbereiten. Am besten 30 Minuten vor dem Treffen die eMeeting -Software testen. Wir müssen aber auch improvisieren können.
	Beobachtung: Es gibt sehr lange Ruhephasen. ~1/3 des Meetings wird geschwiegen. **Emotion:** Die Meetings verlängern sich unnötig. Viel Zeit geht verloren. **Schlussfolgerung:** Besser Leute ansprechen, statt auf Antworten zu warten.
	Beobachtung: eMeeting mit dem LS Tool ist problematisch **Emotion:** Nach ~1h im Projekt ist das Tool-Env. noch völlig unklar. Was hat denn FIN/RUS gemacht? **Schlussfolgerung:**
	Beobachtung: Es gibt sehr lange Ruhephasen. ~1/3 des Meetings wird geschwiegen. Die Meetings verlängern sich unnötig. Viel Zeit geht verloren. **Emotion:** Das ist sehr frustrierend und nicht nur für mich. Auch peinlich. **Schlussfolgerung:** Wir als Team sollten besser aktiver agieren und die Leute selbst ansprechen, statt auf Antworten zu warten.

Vielen Dank für Deine Zeit!

H. Questionnaires for the Evaluation of AnnoEx

This appendix presents the questionnaires (in German) distributed in the evaluation of AnnoEx.

H.1. Questionnaire for the Academic Evaluation of AnnoEx

This appendix presents the questionnaire (in German) distributed in the academic evaluation of AnnoEx. The control group received the same questionnaire but with a different order of questions in the main parts ("Hauptteil 1" and "Hauptteil 2").

(P1v2) Evaluation über die Arbeitsweise mit Dokumenten

Willkommen bei der Evaluation über die Arbeitsweise mit Dokumenten.

Diese Umfrage besteht aus 4 Teilen (=Seiten):

- Allgemeine Fragen,
- Hauptteil 1,
- Hauptteil 2 und
- Abschließende Fragen.

Die Bearbeitung der Fragen kann insgesamt etwa 30 Minuten in Anspruch nehmen. Sie können die Umfrage aber zwischendurch speichern und später fortfahren. Dies können Sie nach jedem Umfrage-Teil tun.

Diese Umfrage enthält 26 Fragen.

Allgemeine Fragen

1 [Allg01]Lesen Sie Literatur, die Sie digital vorfinden, am Bildschirm oder ausgedruckt? *

Bitte wählen Sie nur eine der folgenden Antworten aus:

○ Immer digital

○ Lange Dokumente drucke ich aus

○ Ich drucke Dokumente zum Lesen immer aus

○ Sonstiges [   ]

Bitte beziehen Sie sich nur auf **arbeitsbezogene Dokumente**.

2 [Allg02]Zu welcher Altersgruppe gehören Sie?

Bitte wählen Sie nur eine der folgenden Antworten aus:

○ 21-30 Jahre

○ 31-45 Jahre

○ älter als 45 Jahre

3 [Allg03]In welchem Bereich arbeiten Sie? *

Bitte wählen Sie nur eine der folgenden Antworten aus:

○ Student(in)

○ Wissenschaftliche(r) Mitareiter(in) mit Promotionsziel

○ Postdoc oder Professor(in)

○ In der Wirtschaft tätig

○ Sonstiges [   ]

4 [Allg04]In welchen 3 wichtigsten Forschungsschwerpunkten arbeiten Sie?

Beantworten Sie diese Frage nur, wenn folgende Bedingungen erfüllt sind:
° ((Allg03.NAOK == "A2" or Allg03.NAOK == "A4"))

Bitte geben Sie Ihre Antwort(en) hier ein:

Thema 1 [   ]

Thema 2 [   ]

Thema 3 [   ]

Es reichen Stichpunkte wie z.B.

- Thema 1: Mobile Usability
- Thema 2: Elder People
- Thema 3: Context Awareness.

Geben Sie bitte ihre Themen so genau an, wie Sie können. **Sie können auch weniger Themen angeben.**

Evaluation Teil 1

5 [Ev01]

In diesem 1. Teil (von 2) der Evaluation werden Ihnen zwei Fragen gestellt, die Sie mit Hilfe des weiter unten verlinkten Papers (es öffnet sich im neuen Fenster) beantworten sollen.

Notieren Sie bitte jeweils den Start- und Endzeitpunkt der Bearbeitung der Fragen.

Achtung: Bitte lesen Sie das Paper noch nicht durch. Tun Sie dies erst, wenn Sie anfangen, die Aufgaben unten zu beantworten. Beantworten Sie bitte die beiden Fragen zu diesem Paper nacheinander und ohne lange Pause dazwischen.

Sie können die Umfrage nach dem Bearbeiten dieser Seite gerne unterbrechen. Zum Unterbrechen der Umfrage klicken Sie auf den Button "Später fortfahren" links unten auf dem Bildschirm. Sie werden aufgefordert, einen Namen, Passwort und E-Mail zu zu angeben. Sie bekommen anschließend eine automatische E-Mail mit den Zugangsdaten zugesendet. Zum Fortführen der Umfrage klicken Sie auf den Link in der automatischen E-Mail. In der dann geöffneten Umfrage klicken Sie auf "Zwischengespeicherte Umfrage laden". Damit das Laden der Umfrage klappt, sollten Sie im Browser den Cache der letzten Bearbeitung löschen.

Bei der Bearbeitung dieser Aufgaben geht es nicht um darum, die Fragen so schnell wie möglich zu beantworten. Lesen Sie den Text gründlich durch.

Sie können die Textpassagen, die die Frage beantworten, aus dem Paper herauskopieren.

Jede Antwort sollte mehr als ein Stichwort enthalten.

Link zum Paper (PDF)

6 [Frage1Begin]Notieren Sie bitte den Zeitpunkt, wenn Sie mit der Bearbeitung der Frage 1 beginnen. *

Bitte geben Sie Ihre Antwort hier ein:

[   ]

Die Zeitangabe sollte in der Form **hh:mm** erfolgen.

7 [Frage1]Frage 1: Zu welchem Thema im Bereich Wissenseigenschaften wurde noch wenig recherchiert? Welche Erkenntnisse gibt es bereits dazu?

Bitte geben Sie Ihre Antwort(en) hier ein:

Wenig recherchiert zu

Erkenntnisse

8 [Frage1End]Notieren Sie bitte den Zeitpunkt, wenn Sie mit der Bearbeitung der Frage1 fertig sind. *

Bitte geben Sie Ihre Antwort hier ein:

Die Zeitangabe sollte in der Form hh:mm erfolgen.

9 [Frage2Begin]Notieren Sie bitte den Zeitpunkt, wenn Sie mit der Bearbeitung der Frage 2 beginnen. *

Bitte geben Sie Ihre Antwort hier ein:

Die Zeitangabe sollte in der Form hh:mm erfolgen.

10 [Frage2]Frage 2: Was wird in dem Paper zu Auswirkungen von Managementeinfluss (Involvierung, Positive Einstellung des Managements) in Bezug auf den Wissenstransfer in Allianzen gesagt?

Bitte geben Sie Ihre Antwort(en) hier ein:

11 [Frage2End]Notieren Sie bitte den Zeitpunkt, wenn Sie mit der Bearbeitung der Frage 2 fertig sind. *

Bitte geben Sie Ihre Antwort hier ein:

Die Zeitangabe sollte in der Form hh:mm erfolgen.

Evaluation Teil 2

12 [Ev201]

In diesem 2. Teil (von 2) der Evaluation werden Ihnen zwei Fragen gestellt, die Sie mit Hilfe der weiter unten verlinkten Webseite (sie öffnet sich im neuen Fenster) beantworten sollen.

Diese Seite verlinkt auf das Paper, das für die Beantwortung der Fragen dient. Sie können das Paper dort als PDF herunterladen (müssen aber nicht), indem Sie das Adobe Acrobat-Symbol anklicken.

Das Paper enthält eine Reihe von <u>Annotationen</u>, die auf der Seite (und auch in der PDF selbst) dargestellt sind. Neben einer Annotation befindet sich immer ein Textausschnitt, der den Kontext der Annotation zeigt. Um diesen zu <u>vergrößern</u>, klicken Sie auf das Bild.

Sie können für die Beantwortung der Fragen die Annotationen zu Hilfe nehmen.

Sie können die Textpassagen, die die Frage beantworten, aus dem Paper oder der Webseite mit Annotationen herauskopieren.

Jede Antwort sollte mehr als ein Stichwort enthalten.

Notieren Sie bitte jeweils den Start- und Endzeitpunkt der Bearbeitung der Fragen. Beantworten Sie bitte die beinen Fragen nacheinander und ohne lange Pause dazwischen.

Link zur Webseite

Ab Firefox 19 werden PDFs im Browser per Default nicht mehr mit dem Adobe Reader geöffnet. Das kann evtl. zu Problemen mit der Anzeige von Annotationen führen. Stellen Sie unter Extras > Einstellungen > Anwendungen die Darstellung von PDF auf Adobe Acrobat um.

13 [Teil2Frage1Begin]

Notieren Sie bitte den Zeitpunkt, wenn Sie mit der Bearbeitung von <u>Frage1 beginnen</u>. *

Bitte geben Sie Ihre Antwort hier ein:

Die Zeitangabe sollte in der Form **hh:mm** erfolgen.

14 [Teil2Frage1]Frage 1: Was sind die wichtigsten Änderungen in Bezug auf Webseitennavigation in dieser Studie im Vergleich zu vorangegangenen Studien?

Bitte geben Sie Ihre Antwort(en) hier ein:

Änderungen

15 [Teil2Frage1End]

Notieren Sie bitte den Zeitpunkt, wenn Sie mit der Bearbeitung von <u>Frage 1 fertig</u> sind. *

Bitte geben Sie Ihre Antwort hier ein:

Die Zeitangabe sollte in der Form **hh:mm** erfolgen.

16 [Teil2Frage2Begin]

Notieren Sie bitte den Zeitpunkt, wenn Sie mit der Bearbeitung von <u>Frage 2</u> <u>beginnen</u>. *

Bitte geben Sie Ihre Antwort hier ein:

Die Zeitangabe sollte in der Form **hh:mm** erfolgen.

17 [Teil2Frage2]Frage 2: Was sagt die Studie über die Anzahl offener Browsertabs? Was sind die Vorteile von mehreren offenen Browsertabs?

Bitte geben Sie Ihre Antwort(en) hier ein:

Informationen zu offenen Browsertabs

Vorteile von mehreren Browsertabs

18 [Teil2Frage2End]

Notieren Sie bitte den Zeitpunkt, wenn Sie mit der Bearbeitung von <u>Frage 2 fertig</u> sind. *

Bitte geben Sie Ihre Antwort hier ein:

Die Zeitangabe sollte in der Form **hh:mm** erfolgen.

19 [Teil2Frage3Begin]

Notieren Sie bitte den Zeitpunkt, wenn Sie mit der Bearbeitung von <u>Frage 3</u> <u>beginnen</u>.

*

Bitte geben Sie Ihre Antwort hier ein:

Die Zeitangabe sollte in der Form **hh:mm** erfolgen.

20 [Teil2Frage3]Frage 3: Was waren die Strategien der Studienteilnehmer, um zu Webseiten zu gelangen, die sie oft aufrufen?

Bitte geben Sie Ihre Antwort hier ein:

21 [Teil2Frage3End]

Notieren Sie bitte den Zeitpunkt, wenn Sie mit der Bearbeitung von <u>Frage 3 fertig</u> sind.

*

Bitte geben Sie Ihre Antwort hier ein:

Die Zeitangabe sollte in der Form **hh:mm** erfolgen.

Abschließende Fragen

22 [Abschl01]Haben Sie bei der Beantwortung der Fragen im Teil 2 mit dem annotierten Dokument auch das Paper selbst angeschaut? *

Bitte wählen Sie nur eine der folgenden Antworten aus:

○ Ja

○ Nein

---

23 [Abschl02]Bitte bewerten Sie die Darstellung der Annotationen auf der Webseite bzgl. Aufgabe 2.

Bitte wählen Sie die zutreffende Antwort für jeden Punkt aus:

	überhaupt nicht	eher nicht	etwas	sehr
hilfreich	○	○	○	○
übersichtlich	○	○	○	○

24 [Abschl02Neg]Was war an der Darstellung der Annotationen für die Beantwortung der Aufgabe 2 nicht hilfreich?

Beantworten Sie diese Frage nur, wenn folgende Bedingungen erfüllt sind:
° ((Abschl02_SQ001.NAOK == "A1" or Abschl02_SQ001.NAOK == "A2" or Abschl02_SQ001.NAOK == "A3"))

Bitte geben Sie Ihre Antwort hier ein:

25 [Abschl02Neg2]Was hat Sie an der Übersichtlichkeit der Darstellung gestört?

Beantworten Sie diese Frage nur, wenn folgende Bedingungen erfüllt sind:
° ((Abschl02_SQ002.NAOK == "A1" or Abschl02_SQ002.NAOK == "A2" or Abschl02_SQ002.NAOK == "A3"))

Bitte geben Sie Ihre Antwort hier ein:

26 [Abschl03]Haben Sie sonstige Kommentare, Kritik oder Anregungen?

Bitte geben Sie Ihre Antwort hier ein:

Ich bedanke mich herzlich für die Teilnahme an der Evaluation!

Anna Averbakh

01.01.1970 – 01:00

Übermittlung Ihres ausgefüllten Fragebogens:
Vielen Dank für die Beantwortung des Fragebogens.

H.2. Questionnaire for the Industrial Evaluation of AnnoEx

This appendix presents the questionnaire (in German) distributed in the industrial evaluation of AnnoEx.

Umfrage zum Annotieren von Projektdokumenten

Herzlich Willkommen bei der Umfrage zum Annotatieren von Projektdokumenten!

In dieser Umfrage möchten wir untersuchen, ob und wie Sie mit Annotationen (z.B. Kommentaren, Textmarkierungen) in arbeitsbezogenen Dokumenten arbeiten.

Wir werden Ihre Antworten nur zu Forschungszwecken und selbstverständlich anonymisiert auswerten.

Diese Umfrage enthält 11 Fragen.

Annotieren von arbeitsbezogenen Dokumenten

[]In welchem Bereich sind Sie tätig?

Bitte wählen Sie nur eine der folgenden Antworten aus:

○ Entwickler

○ Analyst

○ Architekt

○ Tester

○ Sonstiges [_____]

[]Welche Arten von Annotationen (digital und in Papierform) begegnen Ihnen in Ihrer Arbeit?

Bitte wählen Sie die zutreffende Antwort für jeden Punkt aus:

	täglich	wöchentlich	montalich	seltener als monatlich	gar nicht
Farbliche Markierungen von Textpassagen	○	○	○	○	○
Textuelle oder bildliche Randnotiz	○	○	○	○	○
Farbliche Textmarkierung mit textueller oder bildlicher Notiz	○	○	○	○	○
Audio- oder Videokommentar	○	○	○	○	○

[]Welche arbeitsbezogenen Dokumente annotieren Sie oder Ihre Kollegen?

Bitte wählen Sie alle zutreffenden Antworten aus:

☐ Fachliche Konzepte

☐ Technische Konzepte

☐ Diagramme

☐ Testfälle

☐ Sonstiges: [_____]

[]Welche Dateitypen annotieren Sie oder Ihre Kollegen?

Bitte wählen Sie die zutreffende Antwort für jeden Punkt aus:

	täglich	wöchentlich	monatlich	seltener als monatlich	gar nicht
Word-Dokumente	○	○	○	○	○
Excel-Dokumente	○	○	○	○	○
PDFs	○	○	○	○	○
Gedruckte Dokumente	○	○	○	○	○

[]Gibt es weitere Dateitypen, die von Ihnen oder Ihren Kollegen annotiert werden und in der vorhergehenden Frage nicht genannt wurden?

Bitte geben Sie Ihre Antwort hier ein:

[   ]

[]Zu welchem Zweck machen Sie oder Ihre Kollegen Annotationen in arbeitsbezogenen Dokumenten?

Bitte wählen Sie alle zutreffenden Antworten aus:

☐ Review-Kommentare beim gemeinsamen Schreiben eines Dokuments

☐ Review-Kommentare bei Korrektur durch Dritte (z.B. Architekturreview, Fachreview)

☐ Gedächtnisstütze, um wichtige Textstellen später schneller zu finden

☐ Zusammenfassungen von Textstellen

☐ Sonstiges: _____

[]Finden Sie das Annotieren von arbeitsbezogenen Dokumenten hilfreich? Bitte begründen Sie Ihre Antwort im Textfeld daneben.

Bitte wählen Sie nur eine der folgenden Antworten aus:

○ sehr hilfreich

○ eher hilfreich

○ eher behindernd

○ sehr behindernd

Bitte schreiben Sie einen Kommentar zu Ihrer Auswahl

[]Wem stellen Sie Ihre annotierten Dokumente zur Verfügung?

Bitte wählen Sie die zutreffende Antwort für jeden Punkt aus:

	ja	nein
nur mir	○	○
auch für enge Kollegen	○	○
auch der Abteilung	○	○
für alle	○	○

[]Wovon hängt es ab, welche Annotationen gespeichert und anderen freigegeben werden sollten?

Bitte wählen Sie alle zutreffenden Antworten aus:

☐ von dem Inhalt der Annotationen und des Dokuments

☐ von der Vertraulichkeit des Dokuments

☐ Sonstiges: _____

[]

Welche Arten von Annotationen sollten Ihrer Meinung nach gespeichert und für spätere Projekte zur Verfügung gestellt werden?

Bitte wählen Sie alle zutreffenden Antworten aus:

☐ Review-Kommentare beim gemeinsamen Schreiben eines Dokuments

☐ Review-Kommentare bei Korrektur durch Dritte (z.B. Architekturreview, Fachreview)

☐ Gedächtnisstütze, um wichtige Textstellen später schneller zu finden

☐ Zusammenfassungen von Textstellen

☐ Sonstiges: _____

[ ]

Wir haben einen Prototypen entwickelt, der in einem Dokument automatisch den markierten Text sowie textuelle Kommentare extrahiert. Die markierten Textpassagen und Kommentare werden jeweils mit einem Ausschnitt (Kontext der Annotation) aus dem Dokument aufbereitet und in einer Wissensdatenbank (z. B. ein Wiki) gespeichert. Die Annotationen werden in dem Wiki suchbar gemacht und verschaffen einen schnellen Überblick über das Dokument.

Ein Beispiel einer solchen Seite für eine PDF-Publikation finden Sie hier:

<u>Link zum Bild (PNG)</u>

Der Link öffnet sich auf einer neuen Seite.

Würde Sie dieses Programm in Ihrer Arbeit unterstützen?

Bitte begründen Sie ihre Antwort kurz.

Bitte wählen Sie nur eine der folgenden Antworten aus:

○ ja

○ nein

Bitte schreiben Sie einen Kommentar zu Ihrer Auswahl

Vielen Dank für Ihre Zeit!

Bei Fragen oder Anregungen kontaktieren Sie bitte

Anna Averbakh

anna.averbakh@inf.uni-hannover.de

http://www.se.uni-hannover.de/pages/de:mitarbeiter_anna_averbakh

Fachgebiet Software Engineering

Institut für praktische Informatik

Leibniz Universität Hannover

Welfengarten 1, Raum G 307

D-30167 Hannover

Tel.:   +49 511 / 762   -19673

Übermittlung Ihres ausgefüllten Fragebogens:
Vielen Dank für die Beantwortung des Fragebogens.

Curriculum Vitae

Personal Information	Anna Averbakh
	born on August 9th 1983 in Leningrad (St. Petersburg), Russia

Education	
01/2010 − 07/2014	Graduate Student in Computer Science, Leibniz Universität Hannover, Germany
10/2007 − 09/2009	Master of Science in Computer Science, Leibniz Universität Hannover, Germany
10/2004 − 09/2007	Bachelor of Science in Computer Science, Leibniz Universität Hannover, Germany
09/1993 − 07/2004	Abitur (general eligibility of university admission), several Elementary Schools and Schillerschule, Hannover, Germany
09/1990 − 08/1993	Elementary School in Leningrad, Russia

Professional Experience	
01/2010 − 06/2014	Researcher at Software Engineering Group, Leibniz Universität Hannover, Germany
01/2008 − 10/2009	Student Assistant at Department *System- und Rechnerarchitektur (SRA)*, Leibniz Universität Hannover, Germany
10/2008 − 01/2009	Internship at Continental AG, Department *Tire Testing Technology, Electronic Equipment*
01/2006 − 06/2008	Student Assistant at the *Division of Computer Graphics*, Leibniz Universität Hannover, Germany

Awards	
2008	Scholarship at Leibniz Universität Hannover

Bibliography

[1] "experience," accessed May 11 2014. [Online]. Available: http://www.oxforddictionaries. com/definition/english/experience?q=experience 8

[2] "How to write in plain English," Accessed 30 July 2014. [Online]. Available: http://www.plainenglish.co.uk/free-guides/60-how-to-write-in-plain-english.html 79

[3] "ISO 9241-11:1998 Ergonomic requirements for office work with visual display terminals (VDTs) – Part 11: Guidance on usability." [Online]. Available: http: //www.iso.org/iso/iso_catalogue/catalogue_tc/catalogue_detail.htm?csnumber=16883 28

[4] "mittlere wortlänge deutsch," accessed Jul 22 2014. [Online]. Available: http: //www.gat-blankenburg.de/pages/fach/info/analyse2.htm 80

[5] "OWL 2 Web Ontology Language," accessed on July 22 2014. [Online]. Available: http://www.w3.org/TR/owl2-syntax/ 11

[6] "Project Management Knowledge - Effort." [Online]. Available: http: //project-management-knowledge.com/definitions/e/effort/ 74

[7] "social network," accessed May 11 2014. [Online]. Available: http: //www.oxforddictionaries.com/definition/english/social-network?q=social+network 25

[8] "IEEE Std 610.12-1990," 2002. 8

[9] C. C. Abt, *Serious Games*. University Press of America, 1987. 150

[10] M. S. Ackerman and T. W. Malone, "Answer Garden: a tool for growing organizational memory," *Proceedings of the conference on Office information systems* -, pp. 31–39, 1990. 50, 60, 61, 62, 69, 171, 193

[11] M. Ackerman, "Augmenting organizational memory: a field study of answer garden," *ACM Transactions on Information Systems (TOIS)*, vol. 16, no. 3, pp. 203–224, Jul. 1998. 171, 182, 185

[12] M. Ackerman and D. McDonald, "Answer Garden 2: Merging Organizational Memory with Collaborative Help," *Computer Supported Cooperative Work*, pp. 97–105, 1996. 60, 62, 171

[13] P. Akhavan, M. Jafari, and M. Fathian, "Exploring the failure factors of implementing knowledge management system in the organizations," *Journal of Knowledge Management Practice*, vol. 6, pp. 1–10, 2005. 21, 161, 162, 163

[14] B. Al-Ani, H. Wilensky, D. Redmiles, and E. Simmons, "An Understanding of the Role of Trust in Knowledge Seeking and Acceptance Practices in Distributed Development Teams," *2011 IEEE Sixth International Conference on Global Software Engineering*, pp. 25–34, Aug. 2011. 161

[15] M. Alavi and D. E. Leidner, "Knowledge management systems: issues, challenges, and benefits," *Communications of the AIS*, vol. 1, no. 2, pp. 1–37, 1999. 1, 161, 162, 164

[16] T. J. Allen, *Managing the Flow of Technology*. MIT Press, Cambridge, 1977. 158

[17] T. Allen, *Managing the Flow of Technology*. Cambridge, MA: MIT Press, 1977. 23

[18] A. Ardichvili, V. Page, and T. Wentling, "Motivation and barriers to participation in virtual knowledge-sharing communities of practice," *Journal of Knowledge Management*, vol. 7, pp. 64–77, 2003. 161, 163

[19] A. Ardichvili, M. Maurer, W. Li, T. Wentling, and R. Stuedemann, "Cultural influences on knowledge sharing through online communities of practice," *Journal of Knowledge Management*, vol. 10, no. 1, pp. 94–107, 2006. 163

[20] G. Attwell and J. Bimrose, "Maturing learning: mashup personal learning environments," *Mash-up Personal Learning Environments (MUPPLE'08)*, 2008. 81

[21] A. Aurum, J. Ross, W. Claes, and M. Handzic, *Managing Software Engineering Knowledge*. Springer-Verlag New York, Inc., Sep. 2003. 45

[22] A. Averbakh, E. Knauss, S. Kiesling, and K. Schneider, "Dedicated Support for Experience Sharing in Distributed Software Projects," *The 26th International Conference on Software Engineering and Knowledge Engineering (SEKE)*, 2014. 80, 129, 130, 131, 134, 135, 137

[23] A. Averbakh, E. Knauss, and O. Liskin, "An Experience Base with Rights Management for Global Software Engineering," *i-KNOW '11 Proceedings of the 11th International*

230

Conference on Knowledge Management and Knowledge Technologies, no. 10, p. 8, 2011. 12, 13, 44, 63, 74, 106, 117, 131, 138, 139, 143, 161, 183

[24] A. Averbakh, K. Niklas, and K. Schneider, "Knowledge from Document Annotations as By-Product in Distributed Software Engineering," *The 26th International Conference on Software Engineering and Knowledge Engineering (SEKE)*, 2014. 80, 116, 122, 124, 127

[25] J. r. P. Bansler and E. C. Havn, "Building community knowledge systems: an empirical study of IT-support for sharing best practices among managers," *Knowledge and Process Management*, vol. 10, no. 3, pp. 156–163, Jul. 2003. 162, 163

[26] F. Barachini, "Cultural and social issues for knowledge sharing," *Journal of Knowledge Management*, vol. 13, no. 1, pp. 98–110, Feb. 2009. 161

[27] R. J. Barson, G. Foster, T. Struck, S. Ratchev, F. Weber, and M. Wunram, "Inter- and Intra-Organisational Barriers to Sharing Knowledge in the Extended," *Proceedings of the eBusiness and eWork*, 2000. 21, 150, 161, 162, 163, 164

[28] C. Bartelt, M. Broy, C. Herrmann, E. Knauss, M. Kuhrmann, A. Rausch, B. Rumpe, and K. Schneider, "Orchestration of Global Software Engineering Projects - Position Paper," *2009 Fourth IEEE International Conference on Global Software Engineering*, pp. 332–337, Jul. 2009. 139

[29] K. M. Bartol and A. Srivastava, "Encouraging Knowledge Sharing: The Role of Organizational Reward Systems," *Journal of Leadership & Organizational Studies*, vol. 9, no. 1, pp. 64–76, Jan. 2002. 149

[30] V. Basili, G. Caldiera, and H. D. Rombach, "The Experience Factory," *Proceedings of the 14th annual Software Engineering Workshop*, vol. 2, pp. 1–19, 1989. 68, 93, 131, 145

[31] V. Basili and P. Costa, "An experience management system for a software engineering research organization," *Software Engineering Workshop*, pp. 29 – 35, 2001. 145

[32] V. Basili, R. Tesoriero, P. Costa, M. Lindvall, I. Rus, F. Shull, and M. Zelkowitz, "Building an Experience Base for Software Engineering : A Report on the First CeBASE eWorkshop," *Product Focused Software Process Improvement*, pp. 110–125, 2001. 50, 57, 58, 67, 69, 145, 178, 191

[33] V. R. Basili, M. Lindvall, and F. Shull, "A Light-Weight Process for Capturing and Evolving Defect Reduction Experience," *Proceedings. Eighth IEEE International Conference on Engineering of Complex Computer Systems*, pp. 129–132, 2002. 1, 50, 57, 58, 69, 93, 145, 162

[34] L. Bean and D. D. Hott, "Wiki: A speedy new tool to manage projects," *Journal of Corporate Accounting & Finance*, vol. 16, no. 5, pp. 3–8, Jul. 2005. 70, 100, 171

[35] T. Bech-Larsen and N. A. Nielsen, "A comparison of five elicitation techniques for elicitation of attributes of low involvement products," *Journal of Economic Psychology*, vol. 20, no. 3, pp. 315–341, Jun. 1999. 149

[36] R. Bergmann, *Experience management: foundations, development methodology, and internet-based applications.* Springer-Verlag Berlin, Heidelberg, 2002. 8, 94

[37] E. Bertino, L. Khan, R. Sandhu, and B. Thuraisingham, "Secure knowledge management: confidentiality, trust, and privacy," *IEEE Transactions on Systems, Man, and Cybernetics - Part A: Systems and Humans*, vol. 36, no. 3, pp. 429–438, May 2006. 29

[38] K.-H. Best, *Satzlängen im Deutschen: Verteilungen, Mittelwerte, Sprachwandel.* Göttinger Beiträge zur Sprachwissenschaft 7, 2002. 79, 80, 131

[39] N. Bettenburg, S. Just, A. Schröter, C. Weiss, R. Premraj, and Thomas Zimmermann, "What makes a good bug report?" *Proceedings of the 16th ACM SIGSOFT International Symposium on Foundations of software engineering*, pp. 308–318, 2008. 129

[40] M. Bohan and A. Chaparro, "To click or not to click: A comparison of two target-selection methods for HCI," *CHI 98 Cconference Summary on Human Computer Interaction*, pp. 219–220, 1998. 80

[41] J. H. Boose, "A survey of knowledge acquisition techniques and tools," *Knowledge Acquisition*, vol. 1, no. 1, pp. 3–37, Mar. 1989. 37

[42] P. Bordia, B. E. Irmer, and D. Abusah, "Differences in sharing knowledge interpersonally and via databases: The role of evaluation apprehension and perceived benefits," *European Journal of Work and Organizational Psychology*, vol. 15, no. 3, pp. 262–280, Sep. 2006. 163

[43] M. Buffa, "Intranet wikis," *Proceedings of the IntraWebs Workshop 2006 at the 15th International World Wide Web Conference*, 2006. 66, 70, 101, 171, 195

[44] W. R. Bukowitz and R. L. Williams, *The knowledge management fieldbook.* Financial Times/Prentice Hall, 2000. 9

[45] V. Bureš, "Cultural barriers in knowledge sharing," *E+ M Ekonomics and Management, Liberec*, vol. 6, 2003. 162, 163

[46] J. Burge, "Knowledge elicitation tool classification," *Artificial Intelligence Research Group, Worcester Polytechnic Institute*, pp. 1–28, 2001. 149

[47] A. M. Burton, N. R. Shadbolt, A. P. Hedgecock, and G. Rugg, "A formal evaluation of knowledge elicitation techniques for expert systems: domain 1," pp. 136–145, Jan. 1988. 149

[48] A. Burton, N. Shadbolt, G. Rugg, and A. Hedgecock, "The efficacy of knowledge elicitation techniques: a comparison across domains and levels of expertise," *Knowledge Acquisition*, pp. 167–178, 1990. 149

[49] A. Cabrera and E. F. Cabrera, "Knowledge-Sharing Dilemmas," *Organization Studies*, vol. 23, no. 5, pp. 687–710, Sep. 2002. 150, 162, 163, 164

[50] E. F. Cabrera and A. Cabrera, "Fostering knowledge sharing through people management practices," *The International Journal of Human Resource Management*, vol. 16, no. 5, pp. 720–735, May 2005. 163, 164

[51] C. Campbell, P. Maglio, A. Cozzi, and B. Dom, "Expertise identification using email communications," *CIKM '03 Proceedings of the twelfth international conference on Information and knowledge management*, pp. 528–531, 2003. 50, 54, 67, 69, 177

[52] H. Cao, V. Govindaraju, and A. Bhardwaj, "Unconstrained handwritten document retrieval," *International Journal on Document Analysis and Recognition (IJDAR)*, vol. 14, no. 2, pp. 145–157, Nov. 2010. 136

[53] P. Carrillo, J. Harding, and A. Choudhary, "Knowledge discovery from post-project reviews," *Construction Management and Economics*, vol. 29, no. 7, pp. 713–723, Jul. 2011. 171

[54] V. Casey and I. Richardson, "Virtual teams: understanding the impact of fear," *Software Process: Improvement and Practice*, pp. 511–526, 2008. 161, 163

[55] G. Casimir, "Knowledge sharing: influences of trust, commitment and cost," *Journal of Knowledge Management*, vol. 16, no. 5, pp. 740–753, 2012. 161, 164

[56] P. Chandler and J. Sweller, "Cognitive Load Theory and the Format of Instruction," *Cognition and Instruction*, vol. 8, no. 4, pp. 293–332, Dec. 1991. 15

[57] C. Chi, M. Zhou, W. Xiao, M. Yang, and E. Wilcox, "Using email to facilitate wiki-based coordinated, collaborative authoring," in *Proceedings of the 2011 annual conference on*

Human factors in computing systems - CHI '11. New York, New York, USA: ACM Press, May 2011, pp. 34–59. 50, 62, 68, 69, 70, 171, 194

[58] C. Chi, M. Zhou, M. Yang, and W. Xiao, "Dandelion: supporting coordinated, collaborative authoring in Wikis," *Proceedings of the SIGCHI Conference on Human Factors in Computing Systems*, pp. 1199–1202, 2010. 50, 62, 69, 183

[59] G. G. Chowdhury, "Natural language processing," *Annual Review of Information Science and Technology*, vol. 37, no. 1, pp. 51–89, Jan. 2005. 41

[60] A. Chua and W. Lam, "Why KM projects fail: a multi-case analysis," *Journal of Knowledge Management*, vol. 9, no. 3, pp. 6–17, Jan. 2005. 161, 162, 163, 164

[61] K. Clark and M. Tomlinson, *The Determinant of Work Effort: Evidence from the Employment in Britain Survey*, 2001. 74

[62] B. Collier, T. DeMarco, and P. Fearey, "A defined process for project post mortem review," *IEEE Software*, vol. 13, no. 4, 1996. 107, 169, 180, 191

[63] J. Conklin, "Designing Organizational Memories: Concept and Method," *Journal of Organizational Computing and Electronic Commerce*, vol. 8, no. 1, pp. 29–55, 2001. 162, 163, 164

[64] C. E. Connelly, D. Zweig, J. Webster, and J. P. Trougakos, "Knowledge hiding in organizations," *Journal of Organizational Behavior*, vol. 1, no. 33, pp. 64–88, 2012. 161

[65] G. Convertino, B. Hanrahan, N. Kong, T. Weksteen, E. Chi, and C. Archambeau, "Mail2Wiki: low-cost sharing and organization on wikis." *Workshop on Collective Intelligence in Organizations: Tools and Studies*, 2010. 183, 184

[66] N. J. Cooke, E. Salas, J. a. Cannon-Bowers, and R. J. Stout, "Measuring Team Knowledge," *Human Factors: The Journal of the Human Factors and Ergonomics Society*, vol. 42, no. 1, pp. 151–173, Mar. 2000. 148

[67] N. Cooke, "Varieties of knowledge elicitation techniques," *International Journal of Human-Computer Studies*, vol. 41, no. 6, pp. 801–849, 1994. 37, 148

[68] A. Cooper, R. Reimann, and D. Cronin, *About face 3: the essentials of interaction design.* John Wiley & Sons, Inc., 2012. 15, 88, 89

[69] M. Cusumano, "How Microsoft makes large teams work like small teams," *Sloan Management Review*, vol. 39, pp. 9–20, 1997. 171, 181

234

[70] M. Cusumano and R. Selby, "How Microsoft builds software," *Communications of the ACM*, vol. 40, no. 6, pp. 53–61, 1997. 171

[71] M. A. Cusumano and R. W. Selby, *Microsoft Secrets: How the World's Most Powerful Software Company Creates Technology, Shapes Markets and Manages People.* The Free Press, 1998. 44, 171, 180, 181

[72] B. Dale and M. Elkjaer, "Fad, fashion and fit: An examination of quality circles, business process re-engineering and statistical process control," *International Journal of Production Economics*, vol. 73, pp. 137–152, 2001. 68

[73] B. Dale and S. Hayward, "Some of the Reasons for Quality Circle Failure: Part I," *Leadership & Organization Development Journal*, vol. 5, no. 1, pp. 11–16, 1984. 68

[74] K. Dalkir, *Knowledge Management in Theory and Practice.* Routledge, 2011. 8, 9, 37, 40, 41, 45, 129

[75] L. Damodaran and W. Olphert, "Barriers and facilitators to the use of knowledge management systems," *Behaviour & Information Technology*, vol. 19, pp. 405–413, 2000. 162, 163, 164

[76] Davenport, Thomas, and L. Prusak, *Working Knowledge: How organizations manage what they know.* Boston, MA: Harvard Business School Press, 1998. 7

[77] T. H. Davenport and G. J. Probst, *Knowledge Management Case Book: Siemens Best Practices.* John Wiley & Sons, Inc., May 2002. 1

[78] D. W. De Long and L. Fahey, "Diagnosing Cultural Barriers to Knowledge Management," *The Academy of Management Executive*, vol. 14, no. 4, pp. 113–127, 2000. 161

[79] D. Dearman, M. Kellar, and K. N. Truong, "An examination of daily information needs and sharing opportunities," *Proceedings of the ACM 2008 conference on Computer supported cooperative work - CSCW '08*, p. 679, 2008. 162

[80] S. Dencheva, C. R. Prause, and W. Prinz, "Dynamic Self-moderation in a Corporate Wiki to Improve Participation and Contribution Quality," *ECSCW 2011: Proceedings of the 12th European Conference on Computer Supported Cooperative Work*, pp. 24–28, 2011. 66, 70, 101, 150, 154, 163, 164, 171, 195

[81] N. Denzin and Y. Lincoln, *Handbook of Qualitative Research.* Newbury Park: Sage Publications, 1984. 31

[82] K. C. Desouza, "Barriers to Effective Use of Knowledge Management Systems in Software Engineering," *Communications of the ACM*, vol. 46, no. 1, pp. 99–101, 2003. 1, 163

[83] K. C. Desouza, Y. Awazu, and P. Baloh, "Managing Knowledge in Global Software Development Efforts: Issues and Practices," *IEEE SOFTWARE*, 2006. 1

[84] S. Dewitte and D. De Cremer, "Self-control and cooperation: different concepts, similar decisions? A question of the right perspective." *The Journal of psychology*, vol. 135, no. 2, pp. 133–53, Mar. 2001. 163

[85] B. Dicicco-Bloom and B. F. Crabtree, "The qualitative research interview." *Medical education*, vol. 40, no. 4, pp. 314–21, Apr. 2006. 81

[86] T. Dingsø yr and R. Conradi, "A survey of case studies of the use of knowledge management in software engineering," *International Journal of Software Engineering and Knowledge Engineering*, vol. 12, no. 4, pp. 391–414, 2002. 1

[87] T. Dingsø yr, N. Moe, and O. y. Nytrø, "Augmenting experience reports with lightweight postmortem reviews," *Product Focused Software Process Improvement*, pp. 167–181, 2001. 50, 59, 60, 69, 98, 106, 146, 169, 171, 179

[88] G. Disterer, "Individual and social barriers to knowledge transfer," *Proceedings of the 34th Hawaii International Conference on System Sciences*, vol. 00, no. c, pp. 1–7, 2001. 150, 162, 163, 164

[89] N. Ducheneaut and V. Bellotti, "E-mail as habitat: an exploration of embedded personal information management," *interactions*, vol. 8, no. 5, pp. 30–38, 2001. 65

[90] A. H. Dutoit, R. McCall, I. Mistrík, and B. Paech, Eds., *Rationale Management in Software Engineering*. Berlin, Heidelberg: Springer Berlin Heidelberg, 2006. 55

[91] C. Dweck, "Motivational processes affecting learning," *American Psychologist*, vol. 41, pp. 1040–1048, 1986. 74

[92] J. H. Dyer and N. W. Hatch, "Relation-specific capabilities and barriers to knowledge transfers: creating advantage through network relationships," *Strategic Management Journal*, vol. 27, no. 8, pp. 701–719, Aug. 2006. 161

[93] S. Easterbrook, S. Janice, S. Margaret-Anne, and D. Daniela, "Selecting Empirical Methods for Software Engineering Research," in *Guide to advanced empirical software engineering*. Springer London, 2008, ch. 11, pp. 285–311. 4

236

[94] J. Edwards, "Managing software engineers and their knowledge," in *Managing Software Engineering Knowledge*, A. Aurum, J. Ross, C. Wohlin, and M. Handzic, Eds., 2003, pp. 5–27. 1

[95] G. Fischer, "Turning breakdowns into opportunities for creativity," *Knowledge-Based Systems*, vol. 7, no. 4, pp. 221–232, Dec. 1994. 44

[96] ——, "Seeding, Evolutionary Growth, and Reseeding: Constructing, Capturing, and Evolving Knowledge in Domain- Oriented Design Environments," *Automated Software Engineering (ASE 1998)*, pp. 1–18, 1998. 26, 97

[97] G. Fischer, A. Lemke, T. Mastaglio, and A. Morch, "Critics: An emerging approach to knowledge-based human-computer interaction," *International Journal of Man-Machine Studies*, vol. 35, no. 5, pp. 695–721, Nov. 1991. 44

[98] G. Fischer, Gerhard Nakakoji, Kumiyo Ostwald, Jonathan Stahl and T. Sumner, "Embedding critics in design environments," *The Knowledge Engineering Review*, vol. 8, no. 04, pp. 285—-307, 1993. 44

[99] S. Fisher and J. Ford, "Differential effects of learner effort and goal orientation on two learning outcomes," *Personnel Psychology*, vol. 51, no. 2, pp. 397–420, Jun. 1998. 74

[100] M. Gagné, "A model of knowledge-sharing motivation," *Human Resource Management*, vol. 48, no. 4, pp. 571–589, 2009. 150

[101] S. Garfield, "Ten reasons why people don't share their knowledge," *Knowledge Management Review*, 2006. 161

[102] T. Gavrilova and T. Andreeva, "Knowledge elicitation techniques in a knowledge management context," *Journal of Knowledge Management*, vol. 16, no. 4, pp. 523–537, Jul. 2012. 148

[103] D. Geer, "Eclipse becomes the dominant Java IDE," *IEEE Computer*, vol. 38, no. 7, pp. 16–18, 2005. 187

[104] R. M. Grant, "Prospering in Dynamically-Competitive Environments: Organizational Capability as Knowledge Integration," *Organization Science*, vol. 7, no. 4, pp. 375–387, 1996. 35

[105] B. Gray, "Informal learning in an online community of practice," *The Journal of Distance Education*, vol. 19, no. 1, 2005. 161, 162, 163, 164

[106] H.-G. Gruber, "Does organisational culture affect the sharing of knowledge?" Dissertation, Carleton University, 2000. 155

[107] T. Gruber, "Ontology," in *Encyclopedia of Database Systems*, L. Liu and M. T. Özsu, Eds. Springer, 2009. 11, 41

[108] J. Grudin, "Social evaluation of the user interface: who does the work and who gets the benefit?" *Proceedings of IFIP INTERACT'87: Human-Computer Interaction*, pp. 805–811, 1987. 146, 164

[109] ——, "Why CSCW applications fail: problems in the design and evaluationof organizational interfaces," *CSCW '88 Proceedings of the 1988 ACM conference on Computer-supported cooperative work*, pp. 85–93, 1988. 2, 27

[110] ——, "Groupware and social dynamics: eight challenges for developers," *Communications of the ACM*, vol. 37, no. 1, pp. 92–105, Jan. 1994. 1, 34, 88, 89, 146, 164

[111] J. Han, M. Kamber, and J. Pei, *Data mining: concepts and techniques*. Morgan Kaufmann, 2006. 41

[112] J. Hannon, "Leveraging HRM to enrich competitive intelligence," *Human Resource Management*, vol. 36, no. 4, pp. 409–422, 1997. 163

[113] B. Hanrahan, G. Bouchard, G. Convertino, T. Weksteen, N. Kong, C. Archambeau, and E. H. Chi, "Mail2Wiki," in *Proceedings of the 5th International Conference on Communities and Technologies - C&T '11*. New York, New York, USA: ACM Press, Jun. 2011, p. 98. 50, 64, 69, 70, 171

[114] B. V. Hanrahan, G. Convertino, and L. Nelson, "Modeling problem difficulty and expertise in stackoverflow," *Proceedings of the ACM 2012 conference on Computer Supported Cooperative Work Companion - CSCW '12*, p. 91, 2012. 61

[115] H.-J. Happel, "Towards Need-driven Knowledge Sharing in Distributed Teams," *I-KNOW '09*, 2009. 161

[116] W. Harrison, "A software engineering lessons learned repository," *Proceedings of the 27 th Annual NASA Goddard / IEEE Software Engineering Workshop (SEW-27'2)*, pp. 139–143, 2003. 170, 171

[117] L. Heeager and P. Nielsen, "Agile Software Development and the Barriers to Transfer of Knowledge: An interpretive Case Study," *Nordic Contributions in IS Research*, vol. 156, pp. 18–39, 2013. 162, 163

238

[118] T. Hellström, U. Malmquist, and J. Mikaelsson, "Decentralizing knowledge Managing knowledge work in a software engineering firm," *The Journal of High Technology Management Research*, vol. 12, no. 1, pp. 25–38, Apr. 2001. 162, 164

[119] J. D. Herbsleb, "Global Software Engineering : The Future of Socio-technical Coordination," *Future of Software Engineering(FOSE'07)*, 2007. 1, 17, 23, 94, 158

[120] J. Herbsleb and A. Mockus, "An empirical study of speed and communication in globally distributed software development," *IEEE Transactions on Software Engineering*, vol. 29, no. 6, pp. 481–494, Jun. 2003. 1, 16, 24, 94, 161

[121] J. Herbsleb and D. Moitra, "Global software development," *Software, IEEE*, pp. 16–20, 2001. 16, 17, 24, 25

[122] J. Herbsleb, D. J. Paulish, and M. Bass, "Global Software Development at Siemens: Experience from Nine Projects," in *Proceedings of the 27th international conference on Software engineering*, ser. ICSE '05.   New York, NY, USA: ACM, 2005, pp. 524–533. 24

[123] J. R. Hobbs, "Granularity," *Proceedings of the Ninth International Joint Conference on Artificial Intelligence*, pp. 432—-435, 1985. 85

[124] R. Hoda, J. Babb, and J. Norbjerg, "Toward Learning Teams," *Software, IEEE*, pp. 1–4, 2013. 163

[125] R. Hoffman and N. Shadbolt, "Eliciting knowledge from experts: A methodological analysis," *Organizational Behavior and Human Decision Processes*, vol. 62, no. 2, pp. 129–158, 1995. 148

[126] G. H. Hofstede, *Culture's consequences: Comparing values, behaviors, institutions and organizations across nations*.   Sage Publications, 2001. 162

[127] J. S. Holste and D. Fields, "Trust and tacit knowledge sharing and use," *Journal of Knowledge Management*, vol. 14, no. 1, pp. 128–140, Feb. 2010. 29, 155, 161

[128] L. J. Holtzblatt, L. E. Damianos, and D. Weiss, "Factors impeding Wiki use in the enterprise: a case study," in *Proceedings of the 28th of the international conference extended abstracts on Human factors in computing systems*, ser. CHI EA '10.   New York, NY, USA: ACM, 2010, pp. 4661–4676. 66, 70, 101, 171, 195

239

[129] B. V. D. Hooff and J. a. D. Ridder, "Knowledge sharing in context: the influence of organizational commitment, communication climate and CMC use on knowledge sharing," *Journal of Knowledge Management*, vol. 8, no. 6, pp. 117–130, 2004. 161

[130] F. Houdek and H. Kempter, "Quality patterns—an approach to packaging software engineering experience," *ACM SIGSOFT Software Engineering Notes*, vol. 22, no. 3, pp. 81–88, May 1997. 172, 175, 185, 195

[131] M. Humayun and G. Cui, "An Empirical Study of the Complex Relationship between KMR and Trust in GSD," *Journal of Software*, vol. 8, no. 4, pp. 776–784, Apr. 2013. 155

[132] K. Husted, "Knowledge-sharing hostility and governance mechanisms: an empirical test," *Journal of Knowledge Management*, vol. 16, no. 5, pp. 754–773, 2012. 155, 161, 163

[133] D. Hutchins, "Quality Circles in Context," *Industrial and Commercial Training*, vol. 15, no. 3, pp. 80–82, 1980. 68

[134] ——, *Quality Circle Handbook*. Pitman Publishing Limited, London, 1986. 68

[135] I. Nonaka, R. Toyama, and N. Konno, "SECI, Ba and Leadership: a Unified Model of Dynamic Knowledge Creation," *Long Range Planning*, vol. 33, pp. 5–34, 2000. 155

[136] S. L. Jarvenpaa and D. E. Leidner, "Communication and Trust in Global Virtual Teams," *Journal of Computer-Mediated Communication*, vol. 3, no. 4, Jun. 2006. 29

[137] S. Jarvenpaa and D. Staples, "The use of collaborative electronic media for information sharing: an exploratory study of determinants," *The Journal of Strategic Information Systems*, vol. 9, no. 2-3, pp. 129–154, Sep. 2000. 161

[138] M. E. Jennex, *Current Issues in Knowledge Management*, M. E. Jennex, Ed. New York, USA: IGI Global, Feb. 2008. 161, 163

[139] M. Jorgensen, "Experience with the accuracy of software maintenance task effort prediction models," *Software Engineering, IEEE Transactions on*, vol. 21, no. 8, 1995. 74

[140] M. E. Kalman, P. Monge, J. Fulk, and R. Heino, "Motivations to Resolve Communication Dilemmas in Database-Mediated Collaboration," *Communication Research*, vol. 29, no. 2, pp. 125–154, Apr. 2002. 163

[141] B. R. Katz and T. J. Allen, "Investigating the Not Invented Here (NIH) svndrome : A look at the performance , tenure , and communication patterns of 50 R & D Project Groups," *R & DManagement*, vol. 12, no. 1, 1982. 163

[142] K. Kautz and V. Mahnke, "Value Creation through IT-supported Knowledge Management? The Utilisation of a Knowledge Management System in a Global Consulting Company," *Journal of Informing Science*, vol. 6, pp. 75–88, 2003. 162, 163, 164

[143] W. Ke and C. Ave, "Critical Factors Affecting the Firm to Share Knowledge with Trading Partners : A Comparative Exploratory Case Study," *ICEC'05*, pp. 177–183, 2005. 161

[144] C. Keet, "A taxonomy of types of granularity," *2006 IEEE International Conference on Granular Computing*, pp. 106–111, 2006. 85

[145] H. Kekre and S. Thepade, "Devnagari Handwritten Character Recognition using LBG vector quantization with gradient masks," *Advances in Technology and Engineering (ICATE), 2013 International Conference on*, pp. 1–4, 2013. 136

[146] G. Keller, M. Nüttgens, and A.-W. Scheer, "Semantische Prozeßmodellierung auf der Grundlage Ereignisgesteuerter Prozeßketten (EPK)," *Publication of Institut für Wirtschaftsinformatik Universität Saarbrücken, Institut für Wirtschaftsinformatik*, vol. 89, 1992. 96

[147] H. Kelley and J. Thibaut, *Interpersonal Relations: A Theory of Independence.* New York: Wiley, 1978. 150

[148] H. Kempter and F. Leippert, "Systematic Software Quality Improvement by Goal- Oriented Measurement and Explicit Reuse of Experience," *BMBF- Statusseminar 1996*, 1996. 50, 63, 131

[149] J. P. Kincaid, R. P. Fishburne, Jr, R. L. Rogers, and B. S. Chissom, "Derivation of New Readability Formulas (Automated Readability Index, Fog Count and Flesch Reading Ease Formula) for Navy Enlisted Personnel," *Tenn: Naval Air Station*, 1975. 131

[150] A. Kini and J. Choobineh, "Trust in electronic commerce: definition and theoretical considerations," in *Proceedings of the Thirty-First Hawaii International Conference on System Sciences*, vol. 4. IEEE Comput. Soc, 1998, pp. 51–61. 29

[151] B. Kitchenham and S. Charters, "Guidelines for performing Systematic Literature Reviews in Software Engineering," Software Engineering Group School of Computer Science and Mathematics Keele University and Department of Computer Science University of Durham Durham, UK, Tech. Rep., 2007. 5, 19, 32, 70

[152] E. Knauss, D. Lübke, and S. Meyer, "Feedback-driven requirements engineering: The Heuristic Requirements Assistant," in *Proceedings of 31st IEEE International Conference*

on *Software Engineering (ICSE '09).*   Vancouver, Canada: IEEE, 2009, pp. 587–590. 131

[153] E. Knauss, "Verbesserung der Dokumentation von Anforderungen auf Basis von Er-fahrungen und Heuristiken," Dissertation, Gottfried Wilhelm Leibniz Universität Han-nover, 2010. 3

[154] D. A. Kolb, *Experiential learning: Experience as the source of learning and development.* Englewood Cliffs, NJ: Prentice-Hall, 1984. 9

[155] U. Koners and K. Goffin, "Learning from Postproject Reviews: A Cross-Case Analysis," *Journal of Product Innovation Management*, vol. 24, no. 3, pp. 242–258, May 2007. 107, 170, 171, 182, 192

[156] B. Kroeber, "Was ist und wozu dient die Normseite." 80

[157] J. Kurman, "Why is Self-Enhancement Low in Certain Collectivist Cultures?: An Investi-gation of Two Competing Explanations," *Journal of Cross-Cultural Psychology*, vol. 34, no. 5, pp. 496–510, Sep. 2003. 163

[158] K. Landerl, "Lesegeschwindigkeitstest (National und International)," in *PISA PLUS 2000*, G. Haider and B. Lang, Eds.   Studien Verlag, 2001, pp. 119–130. 81

[159] T. J. Larsen and J. D. Naumann, "An experimental comparison of abstract and concrete representations in systems analysis," *Information & Management*, vol. 22, no. 1, pp. 29–40, Jan. 1992. 149

[160] S. Laviosa, "Core patterns of lexical use in a comparable corpus of English narrative prose," *Meta: Translators' Journal*, vol. 43, pp. 557–570, 1998. 79

[161] B. Leuf and W. Cunningham, *The Wiki Way: Quick Collaboration on the Web.*   Addison-Wesley Professional, 2001. 70, 100, 171

[162] D. Lübke and E. Knauss, "Dealing with User Requirements and Feedback in SOA Projects," *Software Engineering Methods for Service-Oriented Architecture 2007 (SEM-SOA 2007)*, p. 16, 2007. 129, 133

[163] W. Maalej and H.-J. Happel, "A Lightweight Approach for Knowledge Sharing in Dis-tributed Software Teams," *PAKM*, pp. 14–25, 2008. 147, 161, 164

[164] W. Maalej, D. Panagiotou, and H.-J. Happel, "Towards Effective Management of Soft-ware Knowledge Exploiting the Semantic Wiki Paradigm." *Software Engineering*, vol. 121, pp. 183–197, 2008. 66, 70, 101, 171, 195

[165] M. Mach and M. Owoc, "Granularity of knowledge from different sources," *Intelligent Information Processing IV*, vol. 288, pp. 50–57, 2008. 85

[166] R. Maier and A. Schmidt, "Characterizing Knowledge Maturing," *4th Conference on Professional Knowledge Management - Experiences and Visions (WM '07)*, 2007. 81, 82

[167] C. Marshall, "Toward an ecology of hypertext annotation," *Proceedings of the ninth ACM conference on Hypertext and hypermedia - HYPERTEXT '98*, pp. 40–49, 1998. 117

[168] S. Marshall, "The Index of Cognitive Activity: measuring cognitive workload," *Proceedings of the IEEE 7th Conference on Human Factors and Power Plants*, 2002. 16, 148

[169] T. Matthews and S. Whittaker, "Collaboration personas: A framework for understanding & designing collaborative workplace tools," *Workshop on Collective Intelligence in Organizations, ACM CSCW*, 2010. 155

[170] G. Matturro and A. Silva, "A Model for Capturing and Managing Software Engineering Knowledge and Experience," *j-jucs*, vol. 16, pp. 479–505, 2010. 50, 56, 57, 68, 69, 98, 178, 190

[171] R. Mayer and R. Moreno, "Nine ways to reduce cognitive load in multimedia learning," *Educational psychologist*, no. December 2013, pp. 37–41, 2010. 15, 88

[172] J. A. McCall, P. K. Richards, and G. F. Walters, "Factors in Software Quality," General Electric Company, Rome Air Development Centre, Griffiss Air Force Base New York, Tech. Rep., 1977. 5

[173] R. McDermott, "Overcoming cultural barriers to sharing knowledge," *Journal of knowledge management*, vol. 5, no. 1, pp. 76–85, 2001. 161, 163

[174] D. McDonald and M. Ackerman, "Expertise recommender: a flexible recommendation system and architecture," *Proceedings of the 2000 ACM conference on Computer supported cooperative work*, pp. 231–240, 2000. 61

[175] M. McElroy, "The knowledge life cycle," *ICM Conference on KM*, 1999. 8

[176] M. McLure Wasko and S. Faraj, ""It is what one does": why people participate and help others in electronic communities of practice," *The Journal of Strategic Information Systems*, vol. 9, no. 2-3, pp. 155–173, Sep. 2000. 150, 163

[177] N. Meshkati and P. A. Hancock, Eds., *Human mental workload.* Elsevier, 2011. 89

[178] M. Meyer and M. Zack, "The design and implementation of information products," *Sloan Management Review*, vol. 3, no. 37, pp. 43–59, 1996. 9

[179] S. Meyer, A. Averbakh, T. Ronneberger, and K. Schneider, "Experiences from Establishing Knowledge Management in a Joint Research Project," *Product-Focused Software Process Improvement*, 2012. 2, 3, 4, 20, 21, 26, 47, 66, 67, 70, 101, 156, 161, 162, 163, 164, 171, 195

[180] D. R. Michael and S. L. Chen, "Serious Games: Games That Educate, Train, and Inform," Jul. 2005. 150

[181] G. A. Miller, "The magical number seven, plus or minus two: some limits on our capacity for processing information." *Psychological Review*, vol. 63, no. 2, pp. 81–97, 1956. 15

[182] A. Mockus and J. Herbsleb, "Challenges of global software development," in *Software Metrics Symposium, 2001. METRICS 2001. Proceedings. Seventh International*. IEEE, 2001, pp. 182–184. 24

[183] N. Moe and D. Šmite, "Understanding lacking trust in global software teams: A multi-case study," *Product-Focused Software Process Improvement*, pp. 20–34, 2007. 16, 17, 18

[184] J. Moore and F. Shipman, "A comparison of questionnaire-based and GUI-based requirements gathering," in *Proceedings ASE 2000. Fifteenth IEEE International Conference on Automated Software Engineering*. IEEE Comput. Soc, 2000, pp. 35–43. 149

[185] M. Myllyaho, O. Salo, and J. Kääriäinen, "A review of small and large post-mortem analysis methods," *ICSSEA 2004*, pp. 1–8, 2004. 77, 98, 169

[186] M. Nakayama, E. Binotto, and B. Pilla, "Trust in Virtual Teams: A Performance Indicator," *IFIP 19th World Computer Congress*, vol. 210, pp. 105–113, 2006. 155

[187] B. A. Nardi, D. J. Schiano, M. Gumbrecht, and L. Swartz, "Why we blog," *Communications of the ACM*, vol. 47, no. 12, p. 41, Dec. 2004. 44

[188] T. T. Nguyen, R. W. Smyth, and G. G. Gable, "Knowledge Management Issues and Practices: A Case Study of a Professional Services Firm," *Fifteenth Australasian Conference on Information Systems*, 2004. 161, 163

[189] J. Nielsen, *Usability Engineering*. Academic Press, 1993. 37, 67, 88

[190] I. Nonaka and H. Takeuchi, *The Knowledge-Creating Company: How Japanese Companies Create the Dynamics of Innovation*. Oxford: University Press, 1995. 9, 11, 161

[191] B. Nonnecke and J. Preece, "Why lurkers lurk," *Americas Conference on Information Systems*, pp. 1–10, 2001. 163

[192] I. Ovsiannikov, M. Arbib, and T. McNeill, "Annotation technology," *International Journal of Human-Computer Studies*, vol. 50, no. 4, pp. 329–362, 1999. 117

[193] V. Pipek, J. Hinrichs, and V. Wulf, "Sharing expertise: challenges for technical support," in *Sharing Expertise-Beyond Knowledge Management*, S. Ackermann, V. Pipek, and V. Wulf, Eds. MIT Press, Cambridge, 2003, pp. 111—-136. 162

[194] M. Polanyi, *The tacit dimension*. Garden City, NY: Doubleday, 1966. 10

[195] E. Ras, G. Avram, P. Waterson, and S. Weibelzahl, "Using Weblogs for Knowledge Sharing and Learning in Information Spaces," *Using Weblogs for Knowledge Sharing and Learning in Information Spaces*, vol. 11, no. 3, pp. 394–409, 2005. 161, 162, 164

[196] A. Riege, "Three-dozen knowledge-sharing barriers managers must consider," *Journal of Knowledge Management*, vol. 9, no. 3, pp. 18–35, 2005. 3, 21, 150, 161, 162, 163, 164

[197] J. C. Rivera-Vazquez, L. V. Ortiz-Fournier, and F. R. Flores, "Overcoming cultural barriers for innovation and knowledge sharing," *Journal of Knowledge Management*, vol. 13, no. 5, pp. 257–270, 2009. 161

[198] C. Robson, *Real word research*. Oxford: Blackwell, 2002. 41

[199] B. Rogers, J. Gung, Y. Qiao, and J. E. Burge, "Exploring techniques for rationale extraction from existing documents," *2012 34th International Conference on Software Engineering (ICSE)*, pp. 1313–1316, Jun. 2012. 47, 48, 50, 55, 67, 69, 197

[200] H. Rollett, *Knowledge management: Processes and technologies*. Springer, 2003. 9

[201] N. Romano and J. N. Jr, "Meeting analysis: Findings from research and practice," *Proceedings of the 34th Hawaii International Conference on System Sciences*, vol. 00, pp. 1–13, 2001. 81

[202] T. Romberg, "Wiquila-a Wiki rich client that mixes well with other sources of software project information," *Proceedings of Wikis for Software Engineering*, 2008. 50, 66, 67, 68, 69, 70, 171

[203] B. Rosen, S. Furst, and R. Blackburn, "Overcoming Barriers to Knowledge Sharing in Virtual Teams," *Organizational Dynamics*, vol. 36, no. 3, pp. 259–273, Jan. 2007. 162, 163, 164

[204] G. Roth and A. Kleiner, "Developing organizational memory through learning histories," *Organizational Dynamics*, vol. 27, no. 2, pp. 43–60, 1998. 107, 170, 175, 181, 182, 192

[205] P. Runeson and M. Höst, "Guidelines for conducting and reporting case study research in software engineering," *Empirical Software Engineering*, vol. 14, no. 2, pp. 131–164, Dec. 2008. 5, 41, 42

[206] I. Rus and M. Lindvall, "Knowledge management in software engineering," *Software, IEEE*, vol. 19, no. 3, pp. 26–38, 2002. 2, 163

[207] R. M. Ryan and E. L. Deci, "Self-determination theory and the facilitation of intrinsic motivation, social development, and well-being." *American Psychologist*, vol. 55, no. 1, pp. 68–78, 2000. 27, 149

[208] T. A. Ryan, *Work and effort.* New York: The Ronald Press Company, 1947. 74

[209] V. Santos, "Knowledge sharing barriers in complex research and development projects: An exploratory study on the perceptions of project managers," *Knowledge and Process Management*, vol. 19, no. 1, pp. 27–38, 2012. 161, 163, 164

[210] M. Schindler and M. J. Eppler, "Harvesting project knowledge: a review of project learning methods and success factors," *International Journal of Project Management*, vol. 21, no. 3, pp. 219–228, Apr. 2003. 98, 170, 181

[211] K. Schneider, "Prototypes as assets, not toys: why and how to extract knowledge from prototypes," *Proceedings of the 18th international conference on Software engineering*, pp. 522–531, 1996. 50, 69

[212] ——, "LIDs: a light-weight approach to experience elicitation and reuse," *Product Focused Software Process Improvement*, pp. 407–424, 2000. 33, 34, 50, 58, 59, 69, 73, 82, 97, 98, 106, 111, 139, 146, 147, 150, 154, 164, 172, 179

[213] ——, "Experience Magnets Attracting Experiences, Not Just Storing Them," *PROFES 2001*, pp. 126–140, 2001. 50, 63, 64, 97, 131

[214] ——, "What to expect from software experience exploitation," *Journal of Universal Computer Science (J.UCS)*, vol. 8, no. 6, pp. 570–580, 2002. 3, 50, 53, 69, 108, 162

[215] ——, "Rationale as a By-Product," in *Rationale Management in Software Engineering*, A. H. M. Dutoit, I. Mistrik, and B. Paech, Eds. Berlin / Heidelberg: Springer, 2006, ch. 4, pp. 91—109. 39, 40, 47, 50, 51, 52, 69, 74, 98, 106, 164, 177, 187, 189

[216] ——, *Experience and Knowledge Management in Software Engineering*. Springer, 2009. 8, 9, 12, 13, 26, 33, 34, 37, 41, 42, 43, 44, 73, 86, 87, 89, 146, 162, 163, 164

[217] K. Schneider, S. Meyer, M. Peters, F. Schliephacke, J. Mörschbach, L. Aguirre, and S. M. De, "Feedback in Context : Supporting the Evolution of IT-Ecosystems 1 Introduction : Evolution in IT Ecosystems," *Product-Focused Software Process Improvement, Lecture Notes in Computer Science*, vol. 6156, pp. 191–205, 2010. 161

[218] K. Schneider and T. Schwinn, "Maturing experience base concepts at DaimlerChrysler," *Software Process Improvement and Practice*, vol. 6, no. 2, pp. 85–96, Jun. 2001. 2, 162, 163, 164

[219] K. Schneider, J.-P. von Hunnius, and V. Basili, "Experience in implementing a learning software organization," *IEEE Software*, vol. 19, no. 3, pp. 46–49, 2002. 1

[220] K. Schneider, J.-P. von Hunnius, and J.-p. V. Hunnius, "Effective experience repositories for software engineering," in *ICSE '03: Proceedings of the 25th International Conference on Software Engineering*, vol. 6. Washington, DC, USA: IEEE Computer Society, 2003, pp. 534–539. 108, 162, 163, 164

[221] D. A. Schön, *The reflective practitioner: How professionals think in action*. Basic books, 1983. 44

[222] P. Schugerl, J. Rilling, and P. Charland, "Mining Bug Repositories–A Quality Assessment," *2008 International Conference on Computational Intelligence for Modelling Control & Automation*, pp. 1105–1110, 2008. 129

[223] H. Schuman and S. Presser, "The Open and Closed Question," *American Sociological Review*, vol. 44, no. 5, pp. 692–712, 1979. 41

[224] R. Scupin, "The KJ Method: A Technique for Analyzing Data Derived from Japanese Ethnology," *Human Organization*, vol. 56, no. 2, pp. 233–237, 1997. 59

[225] A. Serenko, "Global ranking of knowledge management and intellectual capital academic journals: 2013 update," *Journal of Knowledge Management*, vol. 17, no. 2, pp. 307–326, 2013. 68

[226] N. Shadbolt and M. Burton, "Knowledge elicitation: a systematic approach," in *Evaluation of human work: a practical ergonomics methodology*, 1995, pp. 406–440. 37, 42

[227] L. Singer and K. Schneider, "It was a bit of a race: Gamification of version control," *2012 Second International Workshop on Games and Software Engineering: Realizing User Engagement with Game Engineering Techniques (GAS)*, pp. 5–8, Jun. 2012. 150

[228] L.-G. Singer, "Improving the Adoption of Software Engineering Practices Through Persuasive Interventions," PhD thesis, Gottfried Wilhelm Leibniz Universität Hannover, 2013. 122, 150, 154

[229] K. Stapel, "Informationsflusstheorie der Softwareentwicklung," Dissertation, Leibniz Universität Hannover, 2012. 7, 14, 18, 189

[230] K. Stapel, E. Knauss, and K. Schneider, "Using FLOW to Improve Communication of Requirements in Globally Distributed Software Projects," *2009 Collaboration and Intercultural Issues on Requirements: Communication, Understanding and Softskills*, pp. 5–14, Aug. 2009. 44, 50, 63, 69, 131, 183

[231] K. Stapel and K. Schneider, "Managing knowledge on communication and information flow in global software projects," *Expert Systems*, pp. 1–35, 2012. 14, 183, 184

[232] R. P. Steel, A. J. Mento, B. L. Dilla, N. K. Ovalle, and R. F. Lloyd, "Factors Influencing the Success and Failure of Two Quality Circle Programs," *Journal of Management*, vol. 11, no. 1, pp. 99–119, Apr. 1985. 68

[233] E. Steinberg, "Cognition and learner control: A literature review," *Journal of Computer-based Instruction*, vol. 16, pp. 117–121, 1989. 74

[234] D. Stenmark, "Leveraging Tacit Organizational Knowledge," *Journal of Management Information Systems*, vol. 17, no. 3, pp. 9–24, 2001. 163

[235] D. Stenmark and R. Lindgren, "Integrating Knowledge Management Systems with Everyday Work : Design Principles Leveraging User Practice," *Proceedings of the 37th Hawaii International Conference on System Sciences*, vol. 00, no. C, pp. 1–9, 2004. 27, 34, 146, 163, 164

[236] J. Storey and E. Barnett, "Knowledge management initiatives: learning from failure," *Journal of Knowledge Management*, vol. 4, no. 2, pp. 145–156, Jan. 2000. 2

[237] P. Y.-T. Sun and J. L. Scott, "An investigation of barriers to knowledge transfer," *Journal of Knowledge Management*, vol. 9, no. 2, pp. 75–90, 2005. 21, 161, 162, 163

[238] T. Susi, M. Johannesson, and P. Backlund, "Serious games: An overview," School of Humanities and Informatics University of Skövde, Sweden, Tech. Rep., 2007. 150

[239] K.-E. Sveiby and R. Simons, "Collaborative climate and effectiveness of knowledge work – an empirical study," *Journal of Knowledge Management*, vol. 6, no. 5, pp. 420–433, Jan. 2002. 161, 163

[240] J. Sweller, "Cognitive load theory, learning difficulty, and instructional design," *Learning and instruction*, vol. 4, pp. 295–312, 1994. 14, 15, 88

[241] G. Szulanski, "Exploring internal stickiness: Impediments to the transfer of best practice within the firm," *Strategic Management Journal*, vol. 17, no. Special Issue: Knowledge and the Firm, pp. 27–43, 1996. 163

[242] S. Tatham, "How to Report Bugs Effectively," accessed Feb 11 2014, 2008. 129

[243] P. Trkman and K. C. Desouza, "Knowledge risks in organizational networks: An exploratory framework," *The Journal of Strategic Information Systems*, vol. 21, no. 1, pp. 1–17, Mar. 2012. 161

[244] R. van Solingen, V. Basili, G. Caldiera, and H. D. Rombach, *Goal Question Metric (GQM) Approach.* John Wiley & Sons, Inc., 2002. 89, 172

[245] T. von Nickisch-Rosenegk, "Unterstützung des Erfahrungsaustauschs mit Hilfe eines Werkzeugs zur Übernahme von PDF- Annotationen in ein Wiki," Bachelorarbeit, Leibniz Universität Hannover, 2012. 117

[246] D. Šmite, C. Wohlin, Z. Galvina, and R. Prikladnicki, *An empirically based terminology and taxonomy for global software engineering*, Jul. 2012, vol. 19, no. 1. 16, 17

[247] D. Šmite, C. Wohlin, T. Gorschek, and R. Feldt, "Empirical evidence in global software engineering: a systematic review," *Empirical Software Engineering*, vol. 15, no. 1, pp. 91–118, Dec. 2009. 1

[248] V. Vuori and J. Okkonen, "Knowledge sharing motivational factors of using an intra-organizational social media platform," *Journal of Knowledge Management*, vol. 16, no. 4, pp. 592–603, Jul. 2012. 161, 162, 163, 164

[249] L. R. Weingart, "Impact of group goals, task component complexity, effort, and planning on group performance," *Journal of Applied Psychology*, vol. 77, no. 5, pp. 682–693, 1992. 74

[250] M. Wendling, "Knowledge sharing barriers in global teams," *Journal of Systems and Information Technology*, vol. 15, no. 3, pp. 239–253, 2013. 24, 161, 162, 163

[251] E. Wenger, "Communities of Practice and Social Learning Systems," *Organization*, vol. 7, no. 2, pp. 225–246, May 2000. 68, 155

[252] K. White, D. Gurzick, and W. Lutters, "Wiki anxiety: impediments to implementing wikis for IT support groups," *Proceedings of the Symposium on Computer Human Interaction for the Management of Information Technology*, pp. 64–67, 2009. 66, 70, 101, 171, 195

[253] R. White, M. Richardson, and Y. Liu, "Effects of community size and contact rate in synchronous social Q&A," *Proceedings of the SIGCHI Conference on Human Factors in Computing Systems*, pp. 2837–2846, 2011. 61

[254] K. M. Wiig, *Knowledge management foundations: thinking about thinking: how people and organizations create, represent, and use knowledge.* Arlington, TX: Schema Press. 8

[255] ——, "Knowledge management: Where did it come from and where will it go?" *Expert Systems with Applications*, vol. 13, no. 1, pp. 1–14, Jul. 1997. 8

[256] W. M. Wilson, L. Rosenberg, and L. E. Hyatt, "Automated Analysis of Requirement Specifications," *International Conference on Software Engineering (ICSE)*, 1997. 129

[257] I. H. Witten and E. Frank, *Data Mining: Practical machine learning tools and techniques.* Morgan Kaufmann, 2005. 41

[258] C. Wohlin, P. Runeson, M. Höst, M. C. Ohlsson, B. Regnell, and A. Wesslén, *Experimentation in Software Engineering: An Introduction.* Kluwer Academic Publishers, 2000. 3, 111

[259] K. Y. Wong, "Critical success factors for implementing knowledge management in small and medium enterprises," in *Industrial Management & Data Systems*, vol. 105, no. 3. Emerald Group Publishing Limited, 2005, pp. 261–279. 162, 163

[260] Y. Ye, Y. Yamamoto, and K. Nakakoji, "A socio-technical framework for supporting programmers," *Proceedings of the the 6th joint meeting of the European software engineering*

conference and the ACM SIGSOFT symposium on The foundations of software engineering, pp. 351–360, 2007. 61

[261] M. H. Zack, *Knowledge and Strategy.*   Woburn: Butterworth Heinemann, 1999. 161

[262] M. V. Zedtwitz, "Organizational learning through post-project reviews in R&D," *R&D Management*, vol. 32, no. 3, pp. 255–268, 2002. 170, 171

[263] S. Zetie, "The quality circle approach to knowledge management," *Managerial Auditing Journal*, vol. 17, no. 6, pp. 317–321, 2002. 68

[264] M.-q. Zhao, Q. Yang, and D.-z. Gao, "Axiomatic Definition of Knowledge Granularity," *RSKT 2008*, no. 1, pp. 348–354, 2008. 85